Advances in Information Systems
and Management Science

Band 65

T0140855

Die Titelgrafik stellt eine Zeichnung aus den Studien konkaver Spiegel verschiedener Krümmungen von Leonardo da Vinci aus dem Jahre 1492 dar. Die zukunftsweisenden, interdisziplinären Gedanken Leonardo da Vincis sind auch für heutige Wissenschaftler ein Leitbild. Er ist Namensgeber für den Leonardo-Campus in Münster, auf dem mehrere Institute unterschiedlicher Fachbereiche und Hochschulen ansässig sind. Die Herausgeber forschen und lehren dort am Institut für Wirtschaftsinformatik im European Research Center for Information Systems (ERCIS).

Advances in Information Systems and Management Science

Band 65

Herausgegeben von

Prof. Dr. Dr. h.c. Dr. h.c. Jörg Becker
Prof. Dr.-Ing. Bernd Hellingrath
Prof. Dr. Stefan Klein
Prof. Dr. Herbert Kuchen
Prof. Dr. Gottfried Vossen

Carolin Wagner

A Process-Centric View on Predictive Maintenance and Fleet Prognostics

Development of a Process Reference Model
and a Development Method for Fleet Prognostics
to Guide Predictive Maintenance Projects

Logos Verlag Berlin

Advances in Information Systems and Management Science

Herausgegeben von

Prof. Dr. Dr. h.c. Dr. h.c. Jörg Becker, Prof. Dr.-Ing. Bernd Hellingrath, Prof. Dr. Stefan Klein,
Prof. Dr. Herbert Kuchen, Prof. Dr. Gottfried Vossen.

Westfälische Wilhelms-Universität Münster
Institut für Wirtschaftsinformatik
Leonardo-Campus 3
D-48149 Münster

Tel.:   +49 (0)251 / 83 - 3 81 00
Fax: +49 (0)251 / 83 - 3 81 09
http://www.wi.uni-muenster.de

D6

Bibliografische Information der Deutschen Nationalbibliothek

Die Deutsche Nationalbibliothek verzeichnet diese Publikation in der Deutschen Nationalbibliografie; detaillierte bibliografische Daten sind im Internet über http://dnb.d-nb.de abrufbar.

ISBN 978-3-8325-5455-2
ISSN 1611-3101

Logos Verlag Berlin GmbH
Georg-Knorr-Str. 4, Geb. 10
12681 Berlin
Tel.:   +49 (0)30 / 42 85 10 90
Fax: +49 (0)30 / 42 85 10 92
http://www.logos-verlag.de

Preface

Predictive maintenance is regarded as one of the central innovations of Industry 4.0. Based on the continuous evaluation of process and machine data, it makes it possible to accurately forecast the remaining lifetime of technical systems and machine components. This enables to determine the best possible time for upcoming maintenance, which can also be integrated smoothly into the planning and control of the production process. Furthermore, predictive maintenance can reduce machine and plant downtimes, cut costs for unplanned breakdowns and reduce spare parts inventories. All these factors lead to an improvement in the profitability of a company applying this maintenance principle.

Despite these advantages of the use of predictive maintenance and its technological implementation, which has been possible for several years, there has not yet been any comprehensive application in industrial practice. Recent surveys addressing plant operators and maintenance service providers show that about 50% of the companies interviewed have implemented initial predictive maintenance projects, but only 4% estimate that the potential has been fully exploited. Reasons for this were seen on the one hand in the non-technical area, where the high implementation effort was described as the greatest challenge. This assessment is especially true for companies that have only limited experience in implementing predictive maintenance. Here, companies lack knowledge about predictive maintenance as well as its structured implementation. If one looks at the previous research efforts in this context, no coherent approach for the structured introduction of this maintenance principle can be found so far. In the technical area, in addition to IT security and infrastructure, the availability and selection of the data required for predictive maintenance and the use of the underlying statistical methods have been cited as further key challenges. With regard to the general problem of available process and machine data for determining the next point of machine failure, one approach currently being considered in research is the use of data not only from one machine but from a fleet of identical or similar machines, which can significantly expand the data basis. However, no classification of the different scenarios of the use of fleet data has been researched yet and no structured approach has been designed. Finally, the construction of an algorithm for predicting the failure time of a plant or component based on fleet data has not been investigated yet.

Ms. Wagner's work addresses this set of complex problems in her thesis with the central question of how the implementation of predictive maintenance projects can be structurally carried out using fleet data. A solution to these problems would on the one side advance research in the area of predictive maintenance significantly and furthermore provide valuable results to be applied in practice.

Ms. Wagner takes on a process view to approach the problems, which is highly useful, especially to enable transferability into practice. The thesis of Ms. Wagner carries three main outcomes.

The first outcome is a reference model for the implementation of predictive maintenance projects. This reference model addresses the problem of the guided implementation of such projects in an excellent way. In addition to the practical applicability, the reference model structures the existing knowledge about the procedure of predictive maintenance for the first time.

Ms. Wagner furthermore developed a method to characterize fleets and the data provided by fleets, breaking new ground in research and practice. With the conceptualization of the term fleet as well as the definition of different fleet types with regard to the intended improvement of the failure prognosis with a conceptual and a data-driven perspective, Ms. Wagner creates an objective basis for further research in this area. The characterization method based on this conceptualization operationalizes the definitional work towards the design of a fleet forecast and goes far beyond the current state of research and practice.

Finally, the method for the development of an algorithm for fleet forecasting as a third outcome derives basic elements and their parameterization for such an algorithm on the basis of the characterization of fleets and fleet data. This is the first research work that structures the existing knowledge toward a methodology for developing an algorithm for fleet forecasting being also operationally feasible.

The results of the work of Ms. Wagner represent a valuable contribution to research in the field of predictive maintenance. It furthermore provides guidance for companies intending to realize a predictive maintenance project and is an excellent example of how research can achieve results for both science and practice. I can therefore highly recommend reading the thesis to both researchers and practitioners.

Prof. Dr.-Ing. Bernd Hellingrath

Acknowledgement

Researching the topic of predictive maintenance and writing a doctoral thesis in this area was a challenging project. This thesis is the result of my work as a research assistant at the Chair for Information Systems and Supply Chain Management at the University of Münster. The PhD board of the School of Business and Economics accepted this doctoral thesis in July 2021. I would like to express my sincere gratitude to the people who have supported me in many ways along the way.

I am especially thankful to my supervisor, Prof. Dr.-Ing. Bernd Hellingrath, who gave me the opportunity to research in the field of predictive maintenance and to write my dissertation. His countless comments and feedback shaped the path of my research and contributed to the success of this thesis. Moreover, I thank Prof. Dr. Fabian Gieseke for co-supervising my thesis as well as Prof. Dr. David Bendig for participating in my disputation panel.

This research would not have been successful without the dedicated participation and contribution of practitioners from industry. Many thanks are thus due to the involved companies and industry experts for their willingness to collaborate on exciting projects on the topic of predictive maintenance and share their valuable insights from the field.

Special thanks go to all my colleagues at the chair. I have greatly appreciated our inspiring discussions and collaborations, as well as the enjoyable after-work activities in Münster and on business and personal trips. Thanks for your continuous encouragement and support. All of you are amazing colleagues and I am fortunate to have been a part of this group.

Most importantly, without my family and friends, none of this would have been possible. I am deeply and sincerely grateful for your strong and continuous support throughout the phases of my PhD. You always showed me what is most important in life and supported me in all my adventures. Great thanks go to my friends for their friendship, helpfulness and at times indulgence along the way. I am forever deeply indebted to my immediate family: my parents for enabling my education and your unconditional support throughout my life as well as my sister and my beloved nieces and nephew, who have been an indispensable pillar during my doctoral studies and have given me great joy. Thank you all sincerely.

Besides my immediate family, this work is dedicated first and foremost to my grandfather, who has shown great appreciation and respect for me taking on this challenge, and by this has given me the strength to complete my dissertation. You are missed.

Carolin Wagner

Table of Contents

List of Figures

List of Abbreviations

ANN	Artificial Neural Network
ANOVA	Analysis of Variance
ARMA	Autoregressive Moving Average Model
C-MAPPS	Commercial Modular Aero-Propulsion System Simulation
CRISP-DM	Cross-Industrial Standard Process for Data Mining
DTC	Diagnostic Trouble Code
DTW	Dynamic-Time Warping
EDR	Edit Distance on Real Sequences
EUC	Euclidean Distance
GPR	Gaussian Process Regression
HI	Health Indicator
HMM	Hidden Markov Model
INCOSE	International Council on Systems Engineering
LCSS	Longest-Common Subsequence
LSTM	Long Short-Term Memory
MAE	Mean Absolute Error
MIPI	Model-based and Integrated Process Improvement
MSE	Mean Squared Error
PCA	Principal Component Analysis
PdM	Predictive Maintenance
PhCP	Phase Coverage Percentage
PHM	Prognostics and Health Management
PMBOK	Project Management Body of Knowledge
PrCP	Process Coverage Percentage
PReMMa	Process Reference Model for Predictive Maintenance
RUL	Remaining Useful Life
RVM	Relevance Vector Machine
SC	Silhouette Coefficient
SVM	Support Vector Machine
SVR	Support Vector Regression

1 Introduction

1.1 Motivation and Problem Statement

Globalization and the intensified competition have fundamentally transformed the business en-vironment (Pawellek 2016, p. 1). As a result, cost and quality pressure and customer orientation have increased. This situation has stimulated the introduction of new technologies and the con-tinuous improvement of required workflows and operations in companies to remain competi-tive (Guillén et al. 2016a, p. 992). This drives today's highly optimized production environments, in which success strongly depends on machine reliability. Unexpected machine failures not only lead to high maintenance costs and downtime but can even cause production targets to be missed. This has given rise to a fundamental change in perspective with regard to maintenance. In the past, maintenance was recognized as a "necessary evil" lacking its own contribution, whereas today, it is an integral part for businesses with high potential for improvements (Levrat et al. 2008, p. 409). The maintenance costs of a complex machine can, in most cases, consume up to half of the initial investment costs by the time it is replaced (Saunders 2007, p. 2; Zaidan 2014, p. 1). Acquisition costs can even be exceeded for well-designed and durable machines (Saunders 2007, p. 2). Conversely, one-third to one-half of the expenses are wasted due to inef-fective maintenance (Zaidan 2014, p. 1; Heng et al. 2009, p. 724). As a result, there is potential for cost-saving that leads to financial benefits, which cannot be addressed adequately by tradi-tional maintenance types, namely corrective and predetermined maintenance. While corrective maintenance is characterized by low machine reliability, predetermined maintenance is charac-terized by high maintenance costs (Haddad et al. 2012, p. 874). This has induced a shift towards more sophisticated maintenance approaches (Bousdekis et al. 2015a, pp. 1226 f.; Jardine et al. 2006, p. 1484).

In the age of digitalization and the fourth industrial revolution, the importance of *predictive mainte-nance* as a proactive maintenance type that anticipates and reduces severe and costly failures has grown significantly in recent years. The availability of machine sensor data incites the realization of predictive maintenance, in which data is used to monitor and analyze the machine's current condition and to predict its future condition. For this purpose, the implementation of predictive maintenance represents a logical step towards maintenance optimization in the industry. Pre-dictive maintenance increases the transparency of the production plant and can, therefore, be used to provide decision-makers with accurate information at the right time (Lee and Bagheri 2015, pp. 299 f.). Using this information, maintenance can be scheduled with minimal interrup-tion of the regular operation (Xue et al. 2008, p. 200).

Predictive maintenance is currently receiving much attention in practice. It is the most fre-quently requested application in the realm of the fourth industrial revolution (Schlick and Negele

2019, p. 32). Thus, it is a topic worth investigating for companies that want to improve them-selves and strive for future success. The implementation of predictive maintenance has created substantial economic benefits for companies that have already successfully incorporated it within their maintenance department (Haarman et al. 2018, p. 19). Several surveys among prac-titioners indicated that the majority are actively engaging with the topic (different surveys report between 60-81% of the addressed participants) (Feldmann et al. 2017, p. 6; Haarman et al. 2018, p. 14). The most frequently mentioned drivers for realizing predictive maintenance are maxim-izing machine uptime and minimizing maintenance costs (Haarman et al. 2018, p. 16). For these two aspects, an optimization potential of 30-50% in machine downtime, as well as 10-40% for after-sales maintenance costs, can be attributed to predictive maintenance (Lucks 2017, p. 781).

Despite these promising facts, predictive maintenance for practical applications is still in its infancy, and the majority of companies are only in the early stages of their implementation (Haarman et al. 2018, p. 5; Schlick and Negele 2019, p. 34). In fact, very few companies have currently integrated and deployed solutions (Feldmann et al. 2017, p. 6). Beyond that, only about 20% of predictive maintenance projects achieve the desired goals (Hart 2017, p. 1). This is due to two prevailing practical challenges in particular. On the one hand, there is a deficiency of knowledge about predictive maintenance and its concrete realization. Up to now, several reali-zations have failed to take a sufficiently systematic approach (Feldmann et al. 2017, p. 13). On the other hand, the lack of high quality and rich data of historical machine failures frequently denotes a limiting factor for the successful implementation of predictive maintenance (Schleichert et al. 2017, p. 11; Blume et al. 2017, p. 6). This may lead to unreliable predictions (Schleichert et al. 2017, p. 11) and thus jeopardizes attaining the desired goal. These two de-scribed current challenges within predictive maintenance have also been stressed several times by the different experts through interviews conducted as part of this thesis. In particular, refer-ring to the first challenge, it was mentioned that lack of knowledge in the area of predictive maintenance and its implementation has often led to misconceptions and resulted in additional effort or even hindered the realization. It was emphasized that the development of a structured process was critical to the company's success. The process enabled a means of communicating the project concept and required steps as well as guidance through the project.

Regarding the research field of predictive maintenance and its related field of prognostics and health management (PHM), it can be generally stated that it features a solid and well-established knowledge base. A considerable amount of the existing literature covers prognostics, which constitutes the critical element in predictive maintenance and focuses on identifying and esti-mating upcoming machine failures (Lee et al. 2017, pp. 9 f.). However, existing research is most times only explanatory, application-dependent, and targets particular problems of the complete implementation process (Elattar et al. 2016, p. 126; Voisin et al. 2010, p. 178; Saxena et al. 2010a, p. 2; Lee et al. 2014b, p. 319). Even though most solutions require a high effort of adaptation and are application-dependent, both the general procedure and the key concepts are universally applicable (IEEE Std 1856-2017, p. 8). Nevertheless, a shortage of and demand for a coherent approach and supportive guidance for the implementation of predictive maintenance has been highlighted by several authors (e.g., Lee et al. 2014b, p. 319; Elattar et al. 2016, p. 126; Tsui et al. 2015, p. 13).

Furthermore, due to the aforementioned data scarcity of real machine data, prognostic models in research are frequently developed on the basis of experimentally generated data, e.g., through simulations or accelerated life testings (Lee et al. 2014a, p. 5; Sutharssan et al. 2015, p. 216). These models, however, over-simplify the actual degradation behavior and are thus often not able to adequately reflect the complexity of real-world systems (Dragomir et al. 2007, p. 435). In particular, this leads to a deficiency of information on the actual degradation behavior of machines and consequently to low robustness under different working conditions and environmental influences (Lee et al. 2014a, p. 5). For this reason, data must be collected over a long time period before accurate predictions can be achieved. Hence, this option is difficult to apply in most industrial implementations. However, since data is often gathered from multiple similar machines, i.e., a fleet of machines, this information could be exploited to expand the available database and increase the data representativeness (Lee et al. 2014a, p. 5; Michau and Fink 2019, p. 1). Nevertheless, this machine fleet data should be evaluated with regard to the individual characteristics and distinct behavior of a single machine. The analysis of fleet data for prognostics is addressed by the field of fleet prognostics. Fleet prognostics is credited with several potential benefits, including reduced uncertainty (Medina-Oliva et al. 2012, p. 2), increased long-term accuracy (Liu et al. 2007, p. 558), as well as robustness (Lee et al. 2014a, p. 5). Nevertheless, to date, the exploitation of fleet data for prognostics remains immature in the research community (Jia et al. 2018, p. 7; Palau et al. 2020, p. 330).

From the above paragraphs, it can be concluded that both identified practical challenges have not yet been comprehensively addressed and studied in the research field of predictive maintenance. In light of these findings, the thesis targets a process-centric view of application-independent predictive maintenance implementations, with a particular emphasis on fleet prognostics. The process-centric view aims to enable systematic design guidance for practitioners and researchers to help overcome the challenges above.

1.2 Research Objective

The previous chapter underlines the importance of supporting companies and researchers in carrying out predictive maintenance, as well as in handling fleet data. In particular, it shows that an existing gap between practice and research must be bridged. This gap can be broken down into two parts. First, it depicts missing systematic guidance for predictive maintenance projects. Second, it addresses the data scarcity problem of real data. For this purpose, it is necessary to exploit the existing knowledge within research, extend it, and make it more accessible for both researchers and practitioners.

With these considerations in mind, the scope of this thesis includes supporting the effective implementation of predictive maintenance, with a particular emphasis on fleet prognostics. It establishes a link between science and industry by preparing and devising theoretical knowledge to overcome practical challenges in predictive maintenance. Consequently, the main research question of this thesis is stated as follows:

> **Main Research Question:** How can the implementation of predictive maintenance projects and fleet prognostics be guided?

From this main research question arises the need for comprehensive and application-independent support for predictive maintenance as well as accompanying assistance for fleet prognostics. In order to derive the envisaged support from the existing knowledge, the theoretical knowledge must be compiled, augmented, and appropriately structured. As the process of a predictive maintenance solution is highly application-specific, support in terms of standardization involves guidelines rather than strict procedures (ISO 13381, p. V). Due to their guiding character, processes are used as a means of structuring. They describe a logical sequence of activities that must be accomplished in order to reach the desired target (Becker et al. 2011, p. 4). Processes are therefore well suited to support the implementation of predictive maintenance and the processing of fleet data for prognostics. The developed processes serve as general guidelines to assist both the structuring of the overall predictive maintenance project as well as the development of prognostic algorithms considering fleet data. Furthermore, these guiding processes can serve all project participants as a basis for discussion and thus, also facilitate the transfer of knowledge and expertise (Wirth and Hipp 2000, p. 2). Consequently, the development of processes with varying abstraction levels is essential to answer the main research question. This leads to the main research goal:

> **Main Research Goal:** Provide a process-centric view of predictive maintenance and fleet prognostics to guide its effective realization.

Three interrelated research questions and goals are formulated in order to address the main research question accordingly. The first part primarily targets the challenge of systematically supporting the implementation of predictive maintenance. This lays the foundation to address the challenge of analyzing fleet data and developing appropriate fleet prognostic algorithms. This is subsequently covered in the second and third part of this thesis. The derived research questions are presented and discussed in the following.

Research Question 1: How can the process of implementing a predictive maintenance project be designed?

Developing a predictive maintenance solution requires a number of steps that are essential for its successful implementation. However, the detailed design of the complete process largely depends on the specific application. Predictive maintenance implementations are targeted by projects due to the limited time duration and unique results (Project Management Institute 2017, p. 4). Thus, in order to support researchers and practitioners during the planning and execution of a predictive maintenance project, the generic implementation process has to be derived. For this purpose, it is necessary to identify the relevant steps and integrate them into an applicable process structure. Due to the strong practical orientation of this thesis, the process development should be driven by real-world requirements. In particular, a comprehensive view of the complete predictive maintenance process is to be devised. It should provide supporting detailed prescriptions for the design of an application-specific process while remaining application-independent. The creation of a generic process model can be effectively approached by means of a process reference model. A process reference model describes recommended practices for a specific area by outlining a set of applicable processes (Rosemann and van der Aalst 2007, p. 2). It is, therefore, driven by the concept of reusability and aims to facilitate the design of application-specific models (Rosemann and van der Aalst 2007, p. 2). Their key desired characteristics are generality, flexibility, completeness, usability, and understandability (Matook and Indulska 2009, pp. 62 f.). As a result, a process reference model for the realization of predictive maintenance enables systematic guidance for researchers and practitioners to define key steps and conceive their unique process tailored to the specific application. For this purpose, the first research question is addressed by the following research goal:

Research Goal 1: Develop a process reference model for the realization of predictive maintenance projects.

Consequently, the result of the first research question constitutes a process reference model for the design of an application-specific process, which covers all relevant phases during the realization of a predictive maintenance project. Besides generic phases, potential detailed processes are elaborated, which can be adapted to specific applications. Above all, the process reference model offers information regarding which processes a predictive maintenance project should cover and therefore describes the highest level of the overall research result. The process reference model aims at high generality and can therefore be applied independently of the consideration of fleet data. The challenge of processing fleet data for prognostics is thus targeted in-depth by the second and third research questions. To this end, a development method for fleet prognostics is derived. Given that the handling of fleet data exclusively affects the algorithm

In summary, this thesis aims to answer the main research question through the provision of a process reference model for predictive maintenance, a characterization method, and a fleet prognostic development method. The three research results require the definition of several processes that target the design of a predictive maintenance solution. The process reference model provides generic processes to be performed in predictive maintenance projects. The process reference model is supplemented by the fleet prognostic development method as part of the second and third research goals. For this reason, the generic algorithm development phase is used as a basis, which is adapted and elaborated for the specific requirements of fleet prognostics. The resulting method facilitates the development of a fleet prognostic algorithm by providing step-by-step guidance for researchers and practitioners. Figure 1 summarizes the three research goals as well as their relations.

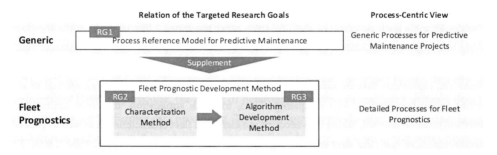

Figure 1 | Summary and Relation of the Research Goals

1.3   Research Design

The stated research objective targets the effective and efficient realization of predictive maintenance and fleet prognostics. Therefore, it is adequately addressed by the design-science research methodology. Unlike natural science, which is a descriptive knowledge-producing activity, design science is prescriptive and knowledge-using (March and Smith 1995, p. 252). It is based on existing theories that are applied, tested, modified, and extended for the specific research objective (Hevner et al. 2004, p. 76). Its purpose is to create, deploy, evaluate and improve a purposeful artifact that can be exploited to effectively and efficiently analyze, design, implement, manage, and use information systems (Hevner et al. 2004). Design science research is embedded between the two pillars, environment and knowledge base, and thus seeks both relevance and rigor. For this purpose, this research methodology lays great importance on the practical significance of the research. In particular, a relevant organizational problem should be focused on, and research should be aligned with the respective business needs (Hevner et al. 2004, p. 79). Furthermore, research rigor is achieved by applying the theoretical foundations during the building of the artifact (Hevner et al. 2004, p. 80). The development process of the artifact comprises an iterative development and evaluation loop (March and Smith 1995, pp. 258–260; Markus et al. 2002, pp. 193–196). The resulting information systems research framework by Hevner et al. (2004) is shown in Figure 2.

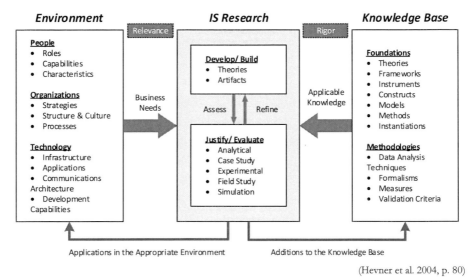

(Hevner et al. 2004, p. 80)

Figure 2 | Integration into the Information Systems Research Framework

Artifacts in design science research can be constructs, models, methods, or instantiations (March and Smith 1995, p. 253). This research contributes to the knowledge base with three artifacts. As the name implies, the first artifact, the process reference model, depicts a *model*. A model in design science research is a representation of the real world with the purpose of understanding the problem and its solution (March and Smith 1995, pp. 256–258; Hevner et al. 2004, pp. 78 f.). It, therefore, provides recommendations for design results (Winter et al. 2009, pp. 7 f.). The developed process reference model illustrates the generic predictive maintenance implementation process. By this, it assists practitioners and researchers in designing their application-specific process. The second and third artifact, the characterization method and the algorithm development method for fleet prognostics, denote *methods*. As already defined in the previous subchapter, methods define processes that prescribe recommendations for activities and provide systematic guidance and instructions for solving the addressed problem (Becker et al. 2007, p. 1; Winter et al. 2009, pp. 7 f.). In this thesis, two multi-step methods are proposed which provide assistance to solve the problem of developing a prognostic algorithm considering fleet data. The two methods together form the fleet prognostic development method.

Design science research must adhere to seven guidelines (Hevner et al. 2004, pp. 82–90). The first guideline, *design as an artifact*, states that the research must address an important organizational problem by means of an artifact. This thesis aims to improve the realization of predictive maintenance and fleet prognostics by means of one model and two methods, as described before. The second guideline, *problem relevance*, calls for the practical orientation of research in the form that a relevant problem for practitioners is solved. Predictive maintenance denotes a currently highly discussed topic in companies. The three artifacts provide guidance for both academics and practitioners. *Design evaluation* is the third guideline. This guideline prescribes the execution of evaluations to verify the utility, quality, and efficacy of the artifacts. The evaluation loop should provide feedback to further enhance the artifact. For this purpose, the thesis applies different types of evaluations. Analytical and empirical evaluation is carried out for the process reference model. The former comprises the descriptive feature-based approach adapted from Fettke and Loos (2003, p. 83), while the latter involves semi-structured expert interviews, which are collected under consideration of the seven stages of interview investigation as defined by Kvale and Brinkmann (2009, p. 102) as well the four interview phases identified by Misoch (2019, p. 68). For the second and third artifact, the characterization method and the algorithm development method, experimental evaluations are performed on the established Commercial Modular Aero-Propulsion System Simulation (C-MAPPS) data set. This data set comprises four data sets of a fleet of turbofan engines generated under different conditions and settings. This allows different fleet characteristics to be covered within the evaluation. To conclude, three application cases are carried out in terms of practical evaluations of the overall research result of the main objective of this thesis. To achieve this, three predictive maintenance projects were realized in close cooperation with various companies. *Research contribution* denotes the fourth guideline, which states that research must enrich the knowledge base either by solving an open problem or by finding a more effective and efficient solution. As described above, the research targets the assistance of predictive maintenance and fleet prognostics, thus providing a more efficient approach to solving this problem. *Research rigor* (fifth guideline) examines the manner in which research is pursued. Rigor in this thesis follows from the exploitation of the relevant knowledge base. It is based on past research, in particular, originating from the fields of process models in predictive maintenance as well as fleet prognostics. This knowledge base is established through systematic literature reviews following the methodology of vom Brocke et al. (2009) and Webster and Watson (2002). In addition, the development and evaluation of the artifacts were accomplished, taking into account existing research development and evaluation methodologies. Subsequently, the *design as a search process* guideline (sixth) claims that the search process is inherently iterative. For the construction of the process reference model, the guideline was implemented by first creating the level of highest abstraction on the basis of the literature. Then, additional levels of detail were added iteratively. Higher abstraction levels are adjusted in case of deviations. For the two methods, the problem is decomposed into simpler and manageable sub-problems. The achieved results are then taken as a starting point, and the overall objective is solved by iteratively combining, adjusting, and extending the component solutions, resulting in a comprehensive characterization process and algorithm development process for fleet prognostics. The design for the three artifacts follows a heuristic search process, which aims at finding a solution which "satisfices" (Simon 1996, p. 27). Lastly, *communication of research* denotes the seventh guideline. Research results should be prepared in a form that is appropriate for the target audience. For this reason, results are structured in a hierarchical manner adressing different audiences. While management at the top level is mainly concerned with the key processes (highest level) for deriving decisions, details are provided at the lower levels for technology-oriented audiences to ease implementation.

The previous paragraph illustrated the grounding of the overall research in design science, particularly in relation to the seven guidelines. Further information about the methodology that was used to develop the respective research results is explained in detail at the beginning of the corresponding chapters. Figure 3 summarizes the key components of the artifact development. Artifacts are designed, taking into account the knowledge base. All three artifacts are evaluated individually by analytical, empirical, and experimental evaluations as part of the build-evaluate loop. The overall thesis result is conclusively validated within three practical projects. The detailed methodological approach for each artifact is presented at the beginning of the corresponding chapters (see Chapters 3.2, 4.2, and 5.2).

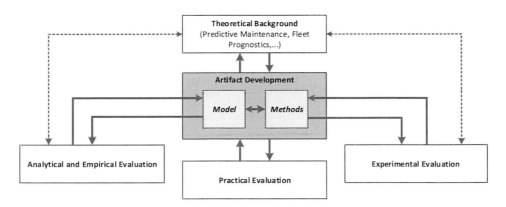

Adapted from Moody and Shanks 2003, p. 647

Figure 3 | Composition of Artifact Development

1.4 Thesis Structure

The structure of the thesis is designed to reflect the presented research objective. The resulting structure is visualized in Figure 4. Up to this point, the current chapter motivates this research and presented the research objective. In addition, the underlying research design is provided. The following chapters pursue the research objective.

Chapter 2 lays the foundation for an informed discussion of the thesis's research objective. To ground the thesis, maintenance management is introduced first, and existing maintenance types are analyzed. Subsequently, predictive maintenance and its related components are addressed in detail, and established data-driven methods are outlined. The chapter concludes with current research gaps in predictive maintenance, which are covered by this thesis.

Chapter 3 targets the development of the process reference model for predictive maintenance. For this purpose, the requirements guiding the development are derived, and the applied methodological approach is presented. This is followed by the discussion of existing process models and related influencing fields. In light of this information, the process reference model is depicted. Lastly, the process reference model is evaluated analytically and empirically.

The process reference model is supplemented by the fleet prognostics method in Chapters 4 and 5. For this purpose, initially, in Chapter 4, the fleet characterization method is inferred. To do so, the research guiding target and the resulting requirements are specified for the fleet characterization method, and the related methodological approach is described. Subsequently, the basis of the method construction is laid by presenting a consolidated view of the current literature. Thereafter, the characterization method is developed, which focuses on an improved conceptualization of the fleet as well as the data-driven classification of fleet prognostic types for the purpose of prognostics. The resulting method is evaluated on the turbofan engine degradation data set. As the last step, current limitations are presented.

Building on these results, the fleet prognostic development method is developed in Chapter 5. Similar to the fleet characterization method, primarily derived requirements, the resulting methodological approach, as well as the targeted knowledge base, are described. This is followed by the presentation of the method, including the fleet approach selection and the generic fleet prognostics implementation processes. To demonstrate the applicability, the method is considered for an exemplary fleet prognostics application, and existing limitations are discussed.

In Chapter 6, the application of the developed artifacts is demonstrated in different industrial projects. Therefore, the three realized predictive maintenance projects are briefly described, analyzed, and discussed in view of the scientific results.

The thesis concludes with a consolidated discussion of the overall result. It summarizes the obtained findings and critically reflects the overarching research. Furthermore, future research opportunities are proposed.

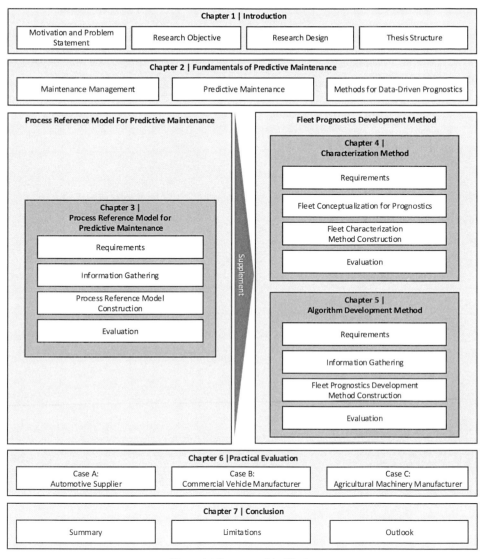

Figure 4 | Structure of the Thesis

2   Fundamentals of Predictive Maintenance

The following chapter introduces the foundations for this thesis's research. It addresses the main concepts in predictive maintenance which are essential for answering the targeted objectives. First, the positioning of maintenance management within spare parts management as well as the different maintenance types, including their key distinctions and application possibilities, are addressed (Chapter 2.1). This is followed by the presentation of predictive maintenance and its key components. Here, the predictive maintenance data analysis process is introduced, and its key steps are subsequently examined (Chapter 2.2). Thereafter, established prognostic methods are investigated. Besides different approach types, data-driven methods are investigated in more depth (Chapter 2.3). In the last chapter (Chapter 2.4), identified research gaps are briefly depicted, and their elaboration in the thesis is outlined.

Since fleet data is still a largely unexplored area that is to be examined in more detail in this thesis, this chapter refrains from introducing fleet data for predictive maintenance in depth. Instead, the foundation for the following research is to be laid here. A detailed discussion of the topic of fleet data for predictive maintenance follows in Chapters 4 and 5.

2.1   Maintenance Management

Predictive maintenance belongs to the field of maintenance management. This chapter, therefore, lays the foundation for maintenance management. For this purpose, initially, the concepts of spare parts management, maintenance management, and maintenance (including the general lifetime failure pattern) are shortly presented (Chapter 2.1.1). Thereafter, the different maintenance types are described and comparatively analyzed (Chapter 2.1.2)

2.1.1 Maintenance Context and Definition

Today, companies face a highly competitive environment and are thus operating under constant cost and efficiency pressure (Pawellek 2016, p. 1). In production, this challenge is often countered by, e.g., automation, inventory reduction, efficient production scheduling, and improved machine design. However, as a result, unexpected disruptions of the production process can hardly be compensated and become thus increasingly expensive. To ensure reliability and continuity of production, improved maintenance management has become essential (Jardine et al. 2006, p. 1484; Pawellek 2016, pp. 1 f.). For this purpose, a shift can be observed in companies that no longer regard maintenance as a necessary expense but instead as a reasonable measure to reduce costs (Khazraei and Deuse 2011, p. 96). This has led to growing attention for maintenance in recent years.

The tasks related to the preservation of the machine's functioning are addressed by spare parts management (Biedermann 2008, p. 6). Spare parts management comprises the ordering and provision of spare parts as well as maintenance control (Strunz 2012, pp. 570 f.). Spare parts logistics and maintenance management are, therefore, integral parts of spare parts management. *Spare parts logistics* support spare parts management by supplying the right spare parts to the right place at the right time with minimal costs (Pawellek 2016, p. 27). In contrast to material logistics, spare parts logistics has to cope with additional challenges, such as higher service requirements, lumpy demand patterns, and a high number of different spare parts (Huiskonen 2001, p. 125; Bacchetti and Saccani 2012, p. 723). *Maintenance management*, on the other side, comprises all activities which specify maintenance objectives, strategies, and responsibilities as well as their implementations (DIN EN 13306, p. 6). For this purpose, all technical, administrative, and managerial measures to preserve the function of a machine are addressed by maintenance (DIN EN 13306, p. 6). Maintenance activities can be grouped into maintenance, inspection, repair, and overhaul (DIN 31051, p. 4). Maintenance denotes the actions to delay the degradation, which is covered by regular maintenance intervals (Khazraei and Deuse 2011, p. 97). Inspection includes activities to identify the current state of the machine, along with the causes of degradation and the definition of possible consequences for future use (DIN 31051, p. 5). In addition, repair covers the physical actions performed to restore the functional state of a faulty machine (DIN 31051, p. 54). Finally, overhaul refers to the combination of measures to increase the properties of a machine (such as reliability, ease of maintenance, and safety) without changing its intended purpose (DIN 31051, p. 6). These maintenance services can be performed either directly by the manufacturer or through maintenance service providers (Bacchetti and Saccani 2012, p. 728).

Maintenance addresses the degradation and failure of machines in order to prevent or counter breakdowns in an appropriate manner. For this purpose, it is important to consider the typical failure patterns over the lifetime with respect to the failure rate, respectively, the hazard rate. This is usually depicted by the Bathtub curve, as shown in Figure 5. Instead of assuming constant failure rates over time, the lifetime is characterized by three stages. In the early stage, the infant mortality region, the failure rate is decreasing. In this stage, failures are usually related to material and manufacturing flaws (Goodman et al. 2019, p. 342). These failures can be eliminated through burn-in testing. This is followed by a stage with constant low random failure. These failures are often caused by fatigue damages (Goodman et al. 2019, p. 343). In the wear-out region, the failure rate is increasing significantly (Vachtsevanos 2006, p. 267). Even though the bathtub curve depicts an often over-simplified representation of the failure rate over time and disrespects usage and environmental conditions (Vachtsevanos 2006, pp. 267 f.), it can be used to gain deeper insights into machine failure rates.

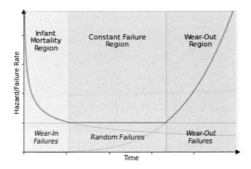

Based on Sikorska et al. 2011, p. 1817; Goodman et al. 2019, p. 343

Figure 5 | Bathtub Curve

2.1.2 Maintenance Types

The increasing importance of maintenance had an essential impact on the determination of the maintenance time from post-failure maintenance to the anticipation of future failures (Gouriveau et al. 2016, p. 3). This paradigm shift has given rise to various types of maintenance. Figure 6 presents a classification of the most commonly described maintenance types.

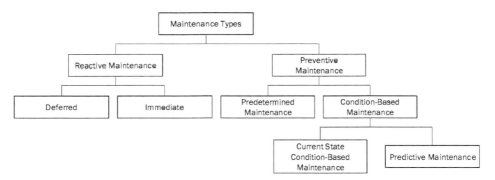

Based on DIN EN 13306, p. 39; Gupta et al. 2012, pp. 2 f.

Figure 6 | Maintenance Types – Overview

On the highest level, maintenance types are categorized into reactive and preventive maintenance. *Reactive maintenance*, also referred to as corrective, unplanned, or run-to-failure maintenance, depicts the oldest maintenance type (Jardine et al. 2006, p. 1484). In this case, machines are operated until failure, and maintenance is only carried out thereafter (Khazraei and Deuse 2011, p. 100). This type can be further divided into deferred (maintenance is delayed according to rules) and immediate (maintenance is carried out without delay) (DIN EN 13306, p. 23). Corrective maintenance allows to utilize the total machine lifetime; however, it is also usually accompanied by longer downtimes due to, e.g., missing spare parts, which can disrupt the production schedule and thus lead to high costs (Lee et al. 2009a, p. 3; Haddad et al. 2012, p. 873).

This maintenance type is, therefore, only suitable for machines that are not critical for the overall production, easy and quickly repairable, and have low investment costs (Gouriveau et al. 2016, p. 4).

The preventive maintenance type counteracts these disadvantages by proactively trying to prevent machine failures (Khazraei and Deuse 2011, p. 98). In this case, maintenance is carried out at pre-specified intervals or according to established criteria (DIN EN 13306, p. 22) to minimize the risk of unexpected machine failure (Gouriveau et al. 2016, p. 4). Preventive maintenance is further split into predetermined maintenance and condition-based maintenance in regards to the information used to plan maintenance.

In *predetermined maintenance*, maintenance is performed in fixed intervals. These intervals can be defined either in relation to time or usage (Haddad et al. 2012, p. 873) with respect to the information provided by the machine manufacturers or previous experiences (Selcuk 2017, p. 2). This maintenance type assumes that machine failures occur periodically according to the Bathtub curve. This, however, might not always be accurate, especially in complex environments where degradation is influenced by a multitude of factors (Kothamasu et al. 2006, pp. 1012 f.). Beyond that, predetermined maintenance can lead to production losses, which are caused by unnecessary maintenance leading to downtime and premature replacement of parts, as well as unexpected catastrophic failures (Lee et al. 2009a, p. 3). Thus, predetermined maintenance offers increased reliability; however, at a substantially higher price due to excessive maintenance (Gouriveau et al. 2016, p. 4).

A further maintenance type that is classified as preventive is condition-based maintenance, which again distinguishes between current state condition-based maintenance and predictive maintenance.[1] *Current state condition-based maintenance*, also referred to as condition monitoring or often only depicted by condition-based maintenance, is characterized by machine condition assessment through monitoring of its performance as well as different parameters (Gouriveau et al. 2016, p. 4). By evaluation of monitoring data, parameter changes in terms of anomalies can be detected (Gouriveau et al. 2016, p. 4). This is possible because the vast majority of machine failures (99%) can be anticipated in advance by means of various indicators (Lee and Wang 1999, p. 7; Bloch and Geitner 2012, p. 305). The collected information can then be used to trigger maintenance with respect to the current machine condition (Jardine et al. 2006, p. 1484). Condition monitoring can either be performed in regular intervals, on request or continuously (DIN EN 13306, pp. 22 f.). Instead of a global approach, in which all machines are treated

---

[1] There is no consensus in the literature on the exact definition and delimitation, so that the terms are often interpreted in various ways. The following classification and naming is therefore based on DIN EN 13306 in accordance with the most common understanding in literature.

equally, in the current state condition-based maintenance, the decisions rely on the actual individual machine health, and maintenance is scheduled only if there is an indication for abnormal behavior (Jardine et al. 2006, p. 1484; Gupta et al. 2012, pp. 3 f.). *Predictive Maintenance* (PdM) goes beyond condition-based maintenance by projecting the monitored health status into the future, which results in an estimate of the expected time of failure (Gouriveau et al. 2016, p. 4). Predictive maintenance, therefore, can be seen as an extension of the current state condition-based maintenance approach (Lee et al. 2009a, p. 3). In addition to estimating the remaining useful life of the machine, predictive maintenance can provide detailed insights on the affected machine, the exact location, the severity as well as the reason for failure (Jardine et al. 2006, pp. 1484 f.; Selcuk 2017, p. 1). Based on this information, proper maintenance can be scheduled at the time when it is needed (Lee et al. 2009a, p. 7). Similar to the current state condition-based maintenance, predictive maintenance targets the individual machine health, but in addition, it also provides an individual failure forecast (Gupta et al. 2012, pp. 3 f.).

Condition-based maintenance and, in particular, predictive maintenance have been the focus of numerous discussions and efforts in recent years, both in industry and academia (Gouriveau et al. 2016, pp. viii–x). This is mainly due to technical advances in the field of sensor technology, which has resulted in greater availability of data (Selcuk 2017, p. 2). Furthermore, the increasing complexity of machines and the growing demand for higher quality and reliability have driven the development of advanced maintenance strategies (Jardine et al. 2006, p. 1484). The benefits of the condition-based maintenance type are manifold. Foremost, condition-based maintenance and, in particular, predictive maintenance can reduce maintenance costs up to 30 % as well as lower the occurrence of machine breakdown by up to 70 % (Sullivan et al. 2010, p. 52). Like predetermined maintenance, this maintenance type reduces the number of failures and, by this, can improve the reliability of the overall production process (Gouriveau et al. 2016, p. 5). In many cases, environmental and safety gains are also attributed (Kothamasu et al. 2006). Besides these advantages, the supply chain can benefit from this advanced maintenance type through enhanced planning, scheduling, and control of operations, as well as optimizing the product and spare parts stock levels (Sun et al. 2012, p. 328). On the downside, however, implementation and support costs are higher as a proper data infrastructure must be established and maintained (Haddad et al. 2012, p. 874). Furthermore, the deterioration of machines must be measurable, and the deterioration rate is required to be low to enable early fault detection, which provides sufficient lead time for maintenance scheduling (Selcuk 2017, p. 3). This maintenance type is particularly recommended for applications in which machine reliability is of utmost importance and involved maintenance costs are extensive (Selcuk 2017, p. 3; Haddad et al. 2012, p. 874). Application areas include power plants, communication systems, and emergency services (Selcuk 2017, p. 1).

Figure 7 compares the described maintenance types with regard to the system health as well as maintenance value. Although the calculation of maintenance value in this context is not clearly specified, it is intended to represent the expected potential for companies in terms of cost and machine reliability (Guillén et al. 2016a, p. 995; Haddad et al. 2012, p. 873). The left figure highlights the point of interaction considering machine health. A significant amount of useful life is lost when maintenance is scheduled in advance (predetermined maintenance). This can be improved through condition-based maintenance. Reactive maintenance consumes the complete useful life. On the other hand, as can be seen in the right figure, the least maintenance value is provided by reactive maintenance due to frequent breakdowns, which involve high costs. If the maintenance interval for predetermined maintenance is not appropriate, this type also has a low value (Haddad et al. 2012, p. 873). In contrast, predictive maintenance has the highest maintenance value, as maintenance can be scheduled prior to breakdown while maximizing the time of machine usage.

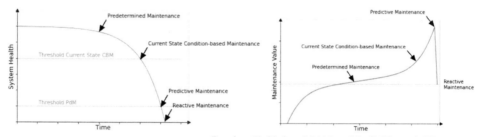

Based on Haddad et al. 2012, p. 873; Guillén et al. 2016a, p. 995

Figure 7 | Value of Maintenance Types

Since machines have different maintenance requirements, *Reliability-Centered Maintenance* provides a framework for determining the individual needs of the machine (Moubray 2001, pp. 1 f.). Reliability-centered maintenance is an overarching maintenance strategy, which supports the assignment of an appropriate maintenance type. This strategy comprises a process to analyze the respective maintenance requirements and relates them to the overall company context in order to establish a viable company maintenance strategy (Sullivan et al. 2010, p. 53). Companies applying reliability-centered maintenance, therefore, often employ a mixture of different maintenance types (Sullivan et al. 2010, p. 53). As a general rule of thumb, reactive maintenance should be used for small parts and equipment, which are not critical and unlikely to fail or which are redundant (NASA 2000, p. 1). Predetermined maintenance is more suitable for types of machines that are subject to wear, with known failure patterns and clear manufacturer recommendations (NASA 2000, p. 1). In particular, machines should have a very similar life span (Guillén et al. 2016a, p. 996). Lastly, in the case of random failure patterns and high criticality, condition-based maintenance and predictive maintenance are most suitable (NASA 2000, p. 1; Sullivan et al. 2010, p. 54).

2.2 Predictive Maintenance

Due to the increasing complexity of machines and the availability of sensor data, predictive maintenance has gained much attention in recent years. As described above, it is the type of maintenance with the highest maintenance value, which, however, requires a meticulous implementation process. The following sub-chapters, therefore, provide a basic knowledge of predictive maintenance and its key components, which are essential for its realization. Chapter 2.2.1 establishes the link to the research field of prognostics and health management and outlines the generic high-level predictive maintenance data analysis process. The elements of the process are examined subsequently, in terms of data acquisition (Chapter 2.2.2), data preparation (Chapter 2.2.3), fault detection and diagnostics (Chapter 2.2.4), prognostics (Chapter 2.2.5), as well as maintenance decision making and presentation (Chapter 2.2.6).

2.2.1 Predictive Maintenance Data Analysis Process

The realization of predictive maintenance entails several steps. Most importantly, the analysis of historical and real-time data is crucial to obtain information for maintenance activities required now or in the future. However, before delving deeper into the data analysis process, it is imperative to first elaborate on the strong interconnections between the research areas of predictive maintenance and Prognostics and Health Management (PHM) as these terms are often used interchangeably in the literature, even though several differences exist. PdM is a maintenance type, which defines when maintenance is carried out, whereas PHM denotes an engineering discipline aiming at maximizing machine utilization and uptime while simultaneously reducing the overall life cycle costs (Vogl et al. 2014a, p. 1). PHM, thus, depicts a key enabler for PdM, which covers methods and technologies to evaluate the reliability of machines in their current life cycle condition (Haddad et al. 2012, p. 873; Pecht 2008, pp. 2 f.). It does not only address industrial maintenance but covers all phases of the machine life cycle, including system design (Sun et al. 2012, p. 324; Gouriveau et al. 2016, pp. 11 f.; Elattar et al. 2016, p. 130). PHM provides useful information to PdM in the form of the remaining useful life, which supports maintenance decision making (Haddad et al. 2012, p. 873; Vogl et al. 2014a, p. 1). While PdM can be implemented without resorting to PHM, e.g., in terms of experience-based prediction, PHM provides the capabilities for real-time prognostics of condition data (Haddad et al. 2012, p. 873; Guillén et al. 2016a, p. 997). PHM, therefore, supports PdM in terms of analyzing available data to schedule maintenance based on current and future machine conditions. Figure 8 provides an overview of the main aspects that differentiate PdM and PHM. Due to the strong interdependence, PdM is used in the following as an umbrella term, while PHM refers exclusively to the data analysis facet within PdM.

	PdM ← Provision of Capabilities	**PHM**
	(Predictive Maintenance)	(Prognostics and Health Management)
Category	Maintenance Type	Engineering Discipline
Goal	Industrial Maintenance Management	Machine Reliability
Focus	Maintenance	All Phases of the Machine Life Cycle

Figure 8 | PdM vs. PHM – Main Distinctions

The integral elements of data analysis within PdM are the detection of emerging faults (fault detection), the identification of the type of fault and failure (diagnosis), and the prediction of the future states (prognostics) (Atamuradov et al. 2017, p. 1). In order to analyze the data with regard to these objectives and to acquire relevant information for machine maintenance, the process for predictive maintenance data analysis can generally be structured into three phases (Jardine et al. 2006, p. 1484). The first phase, data acquisition, aims at obtaining system health data. This data is further analyzed within the second phase, the data processing. Given these results, maintenance decisions are drawn and recommended. Figure 9 provides an overview of the overall PdM data analysis process, including its main phases and steps. Even though not depicted, the overall process is usually cyclic (Gouriveau et al. 2016, p. 12). A brief overview of the steps involved is outlined in the following.

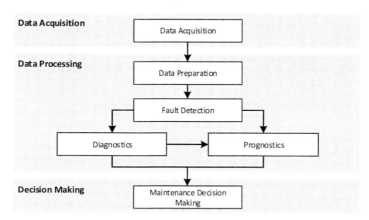

Based on ISO 13374, pp. 1–3; Zaidan 2014, p. 21; Jardine et al. 2006, p. 1484; Gouriveau et al. 2016, pp. 11 f.

Figure 9 | The PdM Data Analysis Process

Data Acquisition | This step collects and stores relevant data from the targeted machine (Jardine et al. 2006, p. 1485). In addition, data on maintenance activities can be added manually by the operator (ISO 13374, p. 1).

Data Preparation | The gathered data is subsequently processed to extract useful information on the condition of the machine, which supports the identification of anomalies as well as the degradation behavior and its evolution over time (Gouriveau et al. 2016, p. 11).

Fault Detection | In this step, data is assessed against the machine's normal condition. In case of deviation from the expected behavior, a fault is present. This is usually done with regard to a pre-defined threshold or based on performance criteria (Gouriveau et al. 2016, p. 11).

Diagnostics | The causes of fault and failure are analyzed within diagnostics by identifying the failure mode and localizing the faulty machine (Gouriveau et al. 2016, p. 11). This can be done after fault detection or post-mortem (Guillén et al. 2016a, p. 994).

Prognostics | This step targets the estimation of future machine health evolution. In general, the current state is identified and extrapolated into the future, also considering future usage (ISO 13374, p. 3). Its key outcome is the remaining useful life of the machine until failure, which provides information for upcoming maintenance requirements.

Maintenance Decision Making | Given the results from the diagnostic and prognostic steps, maintenance decision-making aims to recommend maintenance actions (Gouriveau et al. 2016, p. 11). Furthermore, usage and operational changes could be addressed to extend the lifetime such that the failure happens at a more opportune time (Xue et al. 2008, p. 200).

The unit of analysis considered in the PdM data analysis process can vary, depending on the size, functionality, and complexity of the machine of interest (IEEE Std 1856-2017, p. 24). This can range from individual machine components to the entire machine. However, separate PdM solutions for individual components can be integrated at the machine level, enabling a holistic machine condition analysis. For reasons of simplicity and clarity and in accordance with various standards (e.g., ISO 13374; ISO 17359), the term "machine" is used in the following to refer to the unit of analysis addressed in the PdM data analysis process, even though several individual PdM component solutions might be required depending on the specific application case.

The seven steps of the PdM data analysis process are introduced and explained in more detail in the following sub-chapters.

2.2.2 Data Acquisition[2]

Data acquisition depicts the process of gathering and storing relevant (sensor) data from the targeted machines (Jardine et al. 2006, p. 1485; Atamuradov et al. 2017, p. 4). For the implementation of predictive maintenance, several different sources of data can be exploited. *Machine condition data* is key to PdM as they provide the basis for diagnostics and prognostics. Different sensors, which are attached to specific parts of the machine, are used to monitor the current health in terms of inner-machine characteristics. These sensors provide real-time information on the degradation and fault progression (Lei et al. 2018, p. 800). Well-known techniques include vibration monitoring, thermography, and oil analysis (tribology) (Mobley 2002, p. 4). Out of these techniques, vibration monitoring is the most important tool for monitoring the mechanical conditions of machinery. Vibration sensors measure the amplitudes of the oscillation (Mobley 2002, p. 5). Another common technique is infrared thermography which measures the emission of infrared energy using, e.g., infrared thermometers or imaging systems (Mobley 2002, p. 106). Oil analysis refers to the bearing-lubrication structures of machinery (Mobley 2002, p. 108). For lubricating oil analysis, oil quality is checked frequently and oil is exchanged if required. The wear particle analysis provides direct information about the wearing condition of the machines by assessing particle shape, composition, size, and quantity (Mobley 2002, p. 109). Other techniques include electrical testing (Mobley 2002, p. 112), sound analysis (Bengtsson et al. 2004, p. 78), or the application of ultrasonic sensors (Jardine et al. 2006, p. 1485). In many cases, more than one sensor monitors the machine's condition. The collection of data from multiple sensors provides two main advantages. First, combining same-source signal data can reduce uncertainty with regard to erroneous measurements. Second, the accuracy is increased using multiple types of sensors, which can measure different aspects of the machinery (Hall and Llinas 1997, p. 6).

While machine condition data is essential for PdM data analysis, further data sources improve the overall performance. The degradation depends on different characteristics of the machine and might change in the course of the machine lifecycle (Voisin et al. 2013, pp. 1 f.). It is therefore important to additionally address *machine-related information*. This data comprises the master data and the operating and maintenance history. From the master data, in particular, age is an important information. Furthermore, the operating history provides information on machine utilization (e.g., load and speed). Over time, machines are exposed to different workloads, which influence the speed of degradation (Liao and Lee 2009, p. 895). Besides considering the individual machine, various *external factors* can have an implication on the current health and degradation

---

[2] The sub-chapter is based on an excerpt from the publication "An Overview of Useful Data and Analyzing Techniques for Improved Multivariate Diagnostics and Prognostics in Condition-Based Maintenance" (2016) by Wagner, Saalmann and Hellingrath, which was published in the proceedings of the Annual Conference of the Prognostics and Health Management Society 2016.

behavior. These external factors of machines are mainly determined by the production environment to which the machine is exposed. In general, two influential factors can be distinguished: climatic conditions and the physical environment (Ghodrati 2005, p. 55). The climatic condition of a machine is influenced by the weather, which impacts the microclimate in the production area. In order to quantify the climatic condition, temperature and humidity of the environment are good indicators (Ghodrati 2005, p. 55). The physical environment, on the other hand, is composed of, e.g., dust, smoke, fumes, and corrosive agents (Ghodrati 2005, p. 55).

Besides the above-described data classes, which have a direct impact on the machine, *complementary data* provide additional information, which does not originate from the machine itself. This could be expert knowledge, which depicts the expertise gained by humans in contact with the machine, i.e., by visual inspection, by paying attention to sounds, by touching the machine, or by transferring experiences gained from other machines. In addition, the data available for predictive maintenance can be increased through information and knowledge obtained from similar machines. In this context, a set of similar machines is denoted as a fleet. While previously discussed data sources center on different data for the observed machine, *fleet data* supplement the data for the machine under consideration with all existing data from similar machines. This is particularly beneficial considering that data for individual machines are often scarce and not fully representative (Michau and Fink 2019, p. 1). The reasons for this data scarcity are manifold. Examples include newly developed systems (Elattar et al. 2016, p. 129), lack of run-to-failure data for critical machines (Coble and Hines Wesley 2011, p. 71), as well as few and rare failures. As a result, the usage of fleet data represents an opportunity to increase the amount of available data for predictive maintenance. In the majority of cases, however, the machines within a fleet are of heterogeneous nature despite their similarity. Hence, this heterogeneity has to be taken into account in the subsequent data analysis. If addressed adequately, fleet data can reduce uncertainties (Pecht and Jaai 2010, p. 320; Medina-Oliva et al. 2012, p. 2) and increase accuracy (Liu et al. 2007, p. 558). Unlike the analysis of data from a single machine, which is already well researched, attempts to process fleet data for predictive maintenance are in their infancy, resulting in an immature field of research (Zhou et al. 2016, p. 13; Jia et al. 2018, p. 7).

Figure 10 summarizes the introduced data, which can be gathered in the data acquisition process to be used within the steps of fault detection, diagnostics, and prognostics.

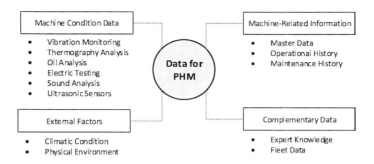

Figure 10 | Overview of Data for PdM

To access the data in the subsequent steps, data needs to be transmitted and stored centrally for subsequent analysis (Lei et al. 2018, pp. 802 f.). For this purpose, maintenance information systems, such as computerized maintenance management systems and enterprise resource planning systems, are available for a collective provision (Davies and Greenough 2000, p. 2; Jardine et al. 2006, p. 1485). Besides automatically gathered data, operators should be able to manually add basic and event data (Jardine et al. 2006, p. 1485).

2.2.3 Data Preparation

Data preparation describes the process that converts raw measurements into informative descriptors and sensor readings (ISO 13374, p. 3). Depending on the following task (fault detection, diagnostics, or prognostics), the data format of the raw measures, as well as the available data (as specified above), the process can vary and needs to be adapted. Figure 11 highlights the key activities of the general data preparation step for PdM. It comprises data cleaning, feature extraction, feature reduction, and selection, as well as health indicator construction.

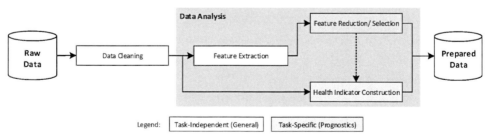

Based on Atamuradov et al. 2017, pp. 4 f.; Gouriveau et al. 2016, pp. 33 f.; Jardine et al. 2006, p. 1486

Figure 11 | General Data Preparation Process

The data acquisition process provides data that is still in a raw format and may be erroneous. *Data cleaning*, thus, aims at improving the data quality by removing data flaws caused, e.g., by sensor faults or manual entries (Jardine et al. 2006, p. 1486). Techniques include missing value imputation, noise reduction through data smoothing, outlier detection and removal, as well as inconsistency elimination (Goebel et al. 2017, p. 25). Beyond that, data fusion, transformation, reduction, and discretization might be required (Goebel et al. 2017, pp. 25 f.).

Data cleaning is followed by a comprehensive and targeted *data analysis* as degradation is often only indirectly observable (Lei et al. 2018, p. 807). Here, relevant and valuable features have been identified that mirror the health condition of the system and provide information for either fault detection, diagnostics, or prognostics (Atamuradov et al. 2017, p. 4). Data analysis includes feature extraction, reduction, and selection, which are often collectively referred to as feature engineering. In addition, most prognostic methods require a well-defined health indicator. The detailed design of the data preparation process is determined by the data type. In general, three data types can be distinguished: value (single values, e.g., temperature data), waveform (signals of time waveform, e.g., vibration data), and multidimensional (multidimensional condition variable, e.g., visual images) (Jardine et al. 2006, p. 1486). For waveform and multidimensional data, *feature extraction* is required (Jardine et al. 2006, p. 1486). For this purpose, methods from the field of signal processing are applied to transform data and extract information in terms of features, which are not directly visible from the raw data (Goebel et al. 2017, p. 26; Gouriveau et al. 2016, p. 35). The definition of features requires expert knowledge and might differ with regard to fault detection, diagnostics, and prognostics (Gouriveau et al. 2016, p. 33). Three common classes of feature extraction techniques exist. The time-domain analysis defines temporal features as statistical parameters of the signal, such as average value or root mean square (Gouriveau et al. 2016, p. 36). These temporal features, however, are very slow with regard to signal changes (Gouriveau et al. 2016, p. 36). This can be counteracted through features defined by frequency-domain analysis. Here, the frequency spectrum is analyzed to identify and isolate faulty frequency levels (Jardine et al. 2006, p. 1487). Time-frequency analysis combines both time and frequency domain and is thus suitable for non-stationary signals (Gouriveau et al. 2016, p. 36). These techniques are, therefore, best suited for prognostics (Jardine et al. 2006, p. 1489). Often too many features are available, which leads to the necessity of *feature reduction* to transfer the available feature space to a lower dimensionality using, for example, principal component analysis or other similar techniques to reduce correlations among features (Gouriveau et al. 2016, p. 48). Furthermore, feature selection can be performed with regard to predefined criteria (Gouriveau et al. 2016, p. 34).

The *construction of a health indicator* (HI) is of particular relevance for prognostics as HIs are designed to directly provide information on the current health (Gouriveau et al. 2016, p. 34). There are two types of health indicators. A physical HI relates to the physics of failure, while a virtual HI denotes an artificially fused signal (Lei et al. 2018, p. 808). These indicators can either be derived from raw data or through feature extraction, selection, and reduction (Gouriveau et al. 2016, p. 64). To evaluate the quality of a health indicator, various measures are proposed. Well-known measures are monotonicity, prognosability, and trendability (Atamuradov et al. 2017,

p. 4). Monotonicity refers to the characteristic that degradation depicts a monotonous increasing or decreasing trend without allowing a system to self-heal (Coble 2010, p. 3). Prognosability analyzes the failure criteria among multiple machines (Coble 2010, p. 44). In the case of high variance between the failure values, the definition of a suitable failure threshold is complicated. Furthermore, trendability assesses the similarity of degradation shape and form among several machines (Coble 2010, p. 44). Beyond that, other measures include robustness, identifiability, and consistency (Lei et al. 2018, pp. 810 f.).

2.2.4 Fault Detection and Diagnostics

The prepared data is processed in fault detection and diagnostics to identify and examine the causes of faults and failures. Here, a *fault* refers to abnormal behavior of a machine, which is characterized by a deviation of an observed variable from a pre-defined tolerable range (Vogl et al. 2016, p. 81), while a *failure* denotes the system condition in which one or more required functions no longer continue to operate (DIN EN 13306, p. 15). In general, the fault precedes the failure and indicates abnormal machine behavior.

The processes of fault detection and diagnostics, often collectively referred to as diagnosis or diagnostics, comprise three main steps, namely detection, isolation, and identification of faults and failures. Detection analyzes the deviation from the normal condition, isolation aims to locate the faulty or failed machine, and identification assesses the nature of fault by assigning a fault type (Jardine et al. 2006, p. 1484). Fault detection triggers both fault diagnostics and prognostics and is therefore often approached separately (Guillén et al. 2016a, p. 993). Beyond that, there are two views on diagnostics depending on the machine's condition. Fault diagnostics is initiated after fault detection and thus carried out before failure. It aims at identifying an incipient failure mode while the machine is still running (Vachtsevanos 2006, p. 176). Failure diagnostics, on the other hand, is performed post-mortem with the objective to improve system maintainability and availability or even feedback information for enhanced system design (Jardine et al. 2006, p. 1491). Failure diagnostics is applied in cases where faults are not predictable (Jardine et al. 2006, p. 1491). The different steps and their timings are presented in Figure 12 as part of an artificial three-stage degradation process.

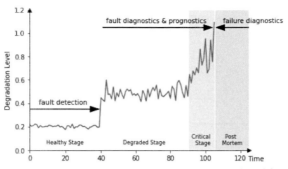

Concept adapted from Lei et al. 2018, p. 813

Figure 12 | Fault Detection and Diagnostics

The process *fault detection*, which is also known under the terms anomaly detection, condition monitoring, and state detection, continuously monitors the machine parameters to identify abnormal behavior as early as possible. This is achieved by comparing the collected data against the baseline parameters (ISO 13379, p. 9). For this purpose, an alarm criterion is set, which is determined either from experiences, manufacturer's specifications, commissioning tests, or by means of statistical data (ISO 17359, p. 8). However, identifying thresholds is a very difficult task and thus often requires in-depth knowledge from practitioners (Atamuradov et al. 2017, p. 5). Beyond that, the operational context, such as the operational condition and environmental effects, have to be factored in the analysis (Guillén et al. 2016a, p. 994).

Diagnostics is based on the fault detection step. This step conducts a thorough examination of the root causes which contributed to the faulty or failure condition (Lee et al. 2014b, p. 314). Diagnostics, therefore, studies the relationship between effects and failure modes (Lee et al. 2014b, p. 315; Guillén et al. 2016a, p. 994). By this, the severity of failure can be quantified (Atamuradov et al. 2017, p. 5). It performs pattern recognition by mapping obtained features to a fault or failure knowledge base (Jardine et al. 2006, pp. 1491 f.). This knowledge base can be established through expert knowledge, physical models, or historical data (Lee et al. 2014b, p. 316).

Methods for solving both fault detection and diagnostics can be classified into knowledge-based and data-driven approaches (ISO 13379, p. 9). Knowledge-based approaches use a-priori knowledge to build a model that explicitly describes the faulty behavior (Katipamula and Brambley 2005, p. 8). They are derived either quantitatively (e.g., physical models) or qualitatively (e.g., rule-based models) (Katipamula and Brambley 2005, p. 8). In contrast, data-driven approaches are trained on historical data and are thus highly automated. This class includes methods, such as classification trees, artificial neural networks, statistical methods, as well as case-based reasoning (ISO 13379, pp. 10–14). Although often no distinction is drawn between methods of fault detection and diagnostics, the objectives of these tasks are different. Fault detection can be achieved through hypothesis testing (Jardine et al. 2006, p. 1492), anomaly detection, or one-/ two-class classifiers. In contrast, diagnostics is associated with pattern recognition and classification. In general, the performance of data-driven methods is measured using false positive and false negative rates (Vachtsevanos 2006, p. 176).

2.2.5 Prognostics

Prognostics is often labeled as the Achilles heel of predictive maintenance (Muller et al. 2008, p. 236). It targets the estimation of the time of failure or any defined future state (Goebel et al. 2017, p. 1; ISO 13381, p. 1). Information gathered through this process provides time to plan maintenance in advance with a high degree of confidence (ISO 13381, p. 4). In general, prognostics can be performed for wear-out failures, whereas it is difficult or hardly possible for random failures (Sun et al. 2010, pp. 3 f.) as depicted in Figure 5. Depending on the application, the term time in prognostics does not necessarily refer to the time scale of minutes and hours but could also be expressed as, e.g., cycles, distances, or liters (Goebel et al. 2017, p. 1).

In the following, first, the different types of prognostics are introduced (Chapter 2.2.5.1). Building on this, degradation-based prognostics is targeted in more depth (Chapter 2.2.5.2).

2.2.5.1 Types of Prognostics

There are various interpretations and understandings of prognostics in literature, ranging from predictions over multiple similar machines using mean statistics to machine-specific predictions based on individual condition data (IEEE Std 1856-2017, p. 10; Goebel et al. 2017, pp. 3 f.). To classify existing approaches, three *types of prognostics* are identified with regard to data being considered. Reliability-based prediction (type I) denotes the simplest approach, which estimates the future failure time based on the average past failures of similar systems (Coble 2010, p. 17). In addition to this purely population-based approach in which each machine is treated in the same manner, stress-based prediction (type II) incorporates past and future unit-specific usage conditions (Coble 2010, p. 19). This prognostics type assumes that different usage behaviors have uneven effects on machine degradation. Ultimately, degradation-based prediction (type III) makes use of machine-specific condition data to capture the machine-specific environment and usage (Coble 2010, p. 25). For this purpose, type-III prognostics analyzes the individual degradation path, which is trended until failure (Coble and Hines Wesley 2008, p. 1). It requires gradual deterioration features that are visible in the data at an early stage (ISO 13381, p. v). The three types of prognostics are tailored towards different stages of the machine's life cycle, as can be seen in Figure 13. At the early stage, where the machine is new and no information is available, the reliability-based prediction is appropriate to get a first idea of the population-wide average lifetime. After the machine has been operated for a while, usage can be further included in the estimate using stress-based prediction methods. If a fault has been detected through fault detection, the degradation-based prediction is applied, taking into account the unit-specific condition data (Coble 2010, pp. 35 f.).

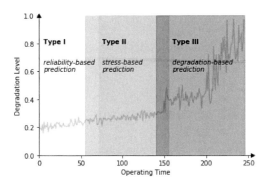

Based on Coble 2010, p. 36

Figure 13 | Types of Prognostics over the Machine's Life Cycle

As an extension to the classification of prognostic types presented above, a fourth type was introduced by Saxena et al. (2014). This type addresses the same stage of the machine's life cycle as type-III prognostics; however, it goes beyond it. Rather than being limited to degradation data from a single machine and the environmental and operating conditions previously encountered, data from multiple similar machines are considered to increase the overall representativeness of the data (Saxena et al. 2014, p. 12). Type-IV, referred to as data analytics-based prediction, aims at identifying complex patterns in unstructured high-dimensional data through discretized prediction (Saxena et al. 2014, p. 4). While this description represents a restricted view of data analysis of multiple similar machines (fleets), a more comprehensive perspective on type-IV is considered in this thesis. The target of type-IV prognostics is thus the consideration of fleet data (as defined in Chapter 2.2.2) for machine-specific degradation modeling. For this type, the machine differences must be factored in and reflected within the individual machine failure prediction. For this reason, this type is renamed *fleet degradation-based prediction,* or in short, *fleet prognostics.* Fleet prognostics has not yet been exhaustively researched and is thus still in its infancy (Jia et al. 2018, p. 7; Palau et al. 2020, p. 2). Yet, various advantages, such as increased (long-term) accuracy (Lee et al. 2015, pp. 20–22) and robustness to external disturbances (Uluyol and Parthasarathy 2012, p. 6; Lapira 2012, pp. 52 f.), have been attributed to fleet prognostics, making it a promising approach for future applications.

The newly defined type-IV can be seen as an extension of type-III and addresses fleet data instead of just individual machine data. However, since research on fleet prognostics is still in its infancy and little grounding research has been done to date, type-III prognostics is needed as a basis for elaborating on the topic and is therefore discussed in detail below. For this pur-pose, in the following, the term prognostics always refers to type-III prognostics, respectively degradation-based prognostics.

2.2.5.2   Degradation-Based Prognostics

The objective of prognostics is the estimation of the remaining useful life (RUL). This measure determines the time left until a failure will occur and can thus be expressed as follows (Jardine et al. 2006, p. 1495):

$$RUL = T - t | T > t, Z(t) \qquad\qquad (2.1)$$

where T defines the random variable of the failure time, t the current time, and $Z(t)$ the past condition profile. In this definition, the RUL represents a random variable that can be described through a probability density function, respectively, a confidence interval (Vogl et al. 2016, p. 82). However, in some cases, the RUL could also be reflected by a point estimate that speci-fies the mean expectation (Jardine et al. 2006, p. 1495).

The realization of prognostics requires four key characteristics on the machine's degradation behavior: (1) machines degrade as a function of use, time, and environmental conditions, (2) degradation is a monotonic process, (3) measurable signs of aging and degradation are visible prior to failure, and (4) there exists a correlation between the degradation model and the meas-urable signs of aging (Uckun et al. 2008, p. 1). Figure 14 depicts a generic degradation behavior, which covers the key characteristics. At the current time, t, the machine has passed the fault threshold. This analysis is usually performed through fault detection. For prognostics, the path is extrapolated or projected into the future (ISO 13381, p. v). As the future degradation path is, however, characterized by uncertainties due to various known and unknown influencing factors, various courses of the path are possible. As degradation progresses over time, various thresholds are exceeded. The alert limit denotes the threshold which is required for appropriate mainte-nance decision making. This limit is usually based on the maintenance lead time (ISO 13381, p. 7). The trip set point denotes the time where the machine is generally stopped to prevent further damages or catastrophic failures (ISO 13381, p. 7). This threshold is often associated with functional failure, where the machine is still operating however not able to fulfill its re-quired function in a satisfactory way (Jardine et al. 2006, pp. 1495 f.) or any other unacceptable behavior (Vogl et al. 2016, p. 82). Lastly, the failure threshold denotes the time when the ma-chine completely fails. The described thresholds can be derived from standards, manufacturers' guidelines, experiences, or historical data (ISO 13381, p. 7). In some cases, thresholds are also modeled under uncertainty, which has to be incorporated in the RUL prediction (Gouriveau et al. 2016, p. 6). As the term remaining useful life implies, it is usually calculated with regard to the trip set point instead of the actual failure time.

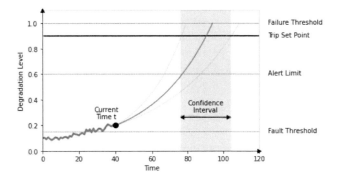

Figure 14 | General Degradation Process

The degradation path can take various forms, which have to be factored in the prognostic analysis. Degradation can be attributed either to normal operations or only initiated after fault detection (Elattar et al. 2016, p. 129). Furthermore, different distinct stages of fault development are possible (Lei et al. 2018, p. 812). For this purpose, the degradation path can be divided into multiple health stages, from healthy to unhealthy, through the usage of clustering approaches (Lei et al. 2018, p. 814). In Figure 14, a two-stage degradation process is depicted. The machine is primarily operating in normal condition until approximately time $t = 20$. This is followed by the degradation stage until the failure threshold is reached. Figure 12, in contrast, depicts a three-stage process, including a healthy, a degraded, and a critical stage. Each of these stages has unique characteristics, which have to be captured in prognostics.

Another important aspect of prognostics is the handling and quantifying of real-world uncertainties, which can cause inaccurate estimates. As there is usually no perfect information available, these uncertainties have to be addressed adequately. Sources of uncertainty can be classified into present state, future, modeling, and prediction method uncertainty (Goebel et al. 2017, pp. 118–120). Present state uncertainty denotes the missing information on the true present state of the machine. This is generally attributed to process and sensor noise and can be covered by filtering techniques (Goebel et al. 2017, p. 118). The uncertainty of future influences, such as environmental effects and operational load, is represented by the second class (Sun et al. 2012, p. 331). A general assumption here is that future influences correspond to historical influences. Model uncertainty depicts the third class. This class accounts for the fact that true behavior is not captured perfectly by the prognostic model (Goebel et al. 2017, p. 119). This often results from model simplifications and parameters (Sun et al. 2012, p. 331). Lastly, the overall uncertainty is usually a combined effect of different sources of uncertainty. This is addressed by the prediction method uncertainty, which represents the combined analysis of uncertainties (Goebel et al. 2017, p. 120). Uncertainties are usually expressed as a failure distribution (in regards to a probability density function) instead of single-point estimates using Bayesian methods or simulation methods (Pecht 2008, p. 56). The further the prognosis is directed into the future, the greater the uncertainty limits (Sun et al. 2012, p. 331). This can be seen in Figure 14, in which the light grey lines denote the upper and lower uncertainty limits of degradation behavior.

Prognostic methods are investigated in-depth in Chapter 2.3, with particular emphasis on data-driven approaches. To assess prognostic methods, a variety of different metrics are available. Besides general application-independent metrics based on accuracy, prediction, and robustness, further prognostics-specific metrics have been developed (Saxena et al. 2008a, pp. 7–15). These specific metrics analyze, e.g., the time at which the prediction first meets the desired accuracy (prediction horizon metric), the percentage of prediction, which falls within pre-defined limits ($\alpha - \gamma$ performance metric) as well as a the improvement of the convergence towards the true value over time (convergence metric) (Saxena et al. 2010a, pp. 13–16).

2.2.6 Maintenance Decision-Making

Fault detection, diagnostics, and prognostics provide information that can be used for improved decision-making in maintenance. Depending on the current health and confidence in the provided information, actions typically include no maintenance action, immediate shutdown, machine inspection, anticipated maintenance scheduling, and operational parameter adjustments (such as load, speed, or throughput) (ISO 17359, pp. 9 f.; Chebel-Morello et al. 2017, pp. 77 f.). Maintenance decision-making in this context refers to a systematic approach that either autonomously determines a suitable action or assists operators in mitigating the effects of machine deterioration (Goebel et al. 2017, p. 253). After the selected maintenance action has been carried out, the activity should be well documented, reviewed, and possibly optimized, as well as fed back to diagnostics and prognostics (ISO 17359, p. 10).

Maintenance decisions serve to optimize the system to maintain it in an operable condition with regard to its functions and state of health (Chebel-Morello et al. 2017, p. 68). As a result, decision-making is often treated as an optimization problem aimed at reducing costs and maximizing the life cycle or operational goal (Chebel-Morello et al. 2017, p. 80; Bougacha et al. 2020, p. 2). These are often NP-complete problems with single- or multi-objective optimization functions (Goebel et al. 2017, pp. 255 f.; Chebel-Morello et al. 2017, p. 80). In general, decision-making is not an isolated process but continuously interacts with previous processes, in particular with prognostics, as decisions impact the system's degradation (Bougacha et al. 2020, p. 2).

The decisions highly depend on the application and the available time horizon. Three general types of decisions are derived by Chebel-Morello et al. (2017, pp. 69–77), namely automatic control, mission optimization, and maintenance optimization. *Automatic control* depicts the decisions which have to be made immediately and are performed directly on the machine's programmable logic controller (small computer enabling the regulation of a machine) (Goebel et al. 2017, p. 256). These decisions include immediate shutdown and, in the case of redundant machines, task reallocation to a different machine with the same functionality. For the latter, a set of redundant machines is required. The performed action then targets a changed configuration to avoid the immediate breakdown of the entire machine (Goebel et al. 2017, pp. 256 f.). Automatic control decisions are often based on real-time and supervised control through fault detection (Bonissone and Iyer 2007, p. 930). The second decision type covers *mission optimization*. A mission in this context refers to a quest or job that must be accomplished by multiple machines in a joint effort (Chebel-Morello et al. 2017, p. 73). In this case, the fulfillment of the mission's objective is optimized (Chebel-Morello et al. 2017, p. 73). This could be reached through replanning or rescheduling activities or adjusting the mission's objective in order to postpone the failure after mission completion (Goebel et al. 2017, p. 26). In case the decision time horizon spans from minutes to days or even longer, *optimized maintenance planning* can be addressed (Chebel-Morello et al. 2017, p. 79). Maintenance optimization is most frequently addressed in the context of post-prognostics decision-making (Chebel-Morello et al. 2017, p. 71). It aims to find the optimal set of maintenance actions under consideration of various trade-offs and constraints (Goebel et al. 2017, pp. 277 f.). For this, further information can be considered, such as operational safety, productivity, and life cycle costs (Goebel et al. 2017, p. 277). Beyond these types, a mixed decision that includes both mission and maintenance optimization is possible (Bougacha et al. 2020, p. 3). Figure 15 highlights the presented decision types with regard to the time horizon and the addressed optimization object. While automatic control is usually performed on the component or machine level, optimized maintenance planning targets the overall machine. Due to the cross-machine objective of a mission, distributed systems are addressed during mission optimization. In general, fault diagnostics and prognostics information contribute to short and medium-term decisions and are therefore mainly targeted at tactical and operational time (Bonissone and Iyer 2007, pp. 930–933).

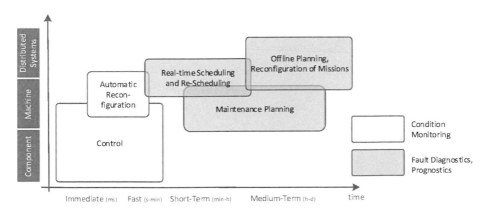

Chebel-Morello et al. 2017, p. 79

Figure 15 | Typology of Maintenance Decision-Making

Lastly, the information obtained from all previous steps must be provided to decision-makers in a suitable format via a human-machine interface (Atamuradov et al. 2017, p. 6; Gouriveau et al. 2016, p. 11).

2.3 Methods for Data-Driven Prognostics

As this thesis considers prognostics in detail within the third research objective, methods for remaining useful life estimation are depicted in the following. For this purpose, initially, different prognostic approaches are portrayed (Chapter 2.3.1) and data-driven methods are introduced (Chapter 2.3.2). Most known data-driven methods for prognostics are depicted thereafter with respect to the classes stochastic methods (Chapter 2.3.3), statistical methods (Chapter 2.3.4), and machine learning methods (Chapter 2.3.5).

For this chapter and the thesis, it is relevant to distinguish the different notions of algorithm, method, technique, and approach. Even though these terms are often used interchangeably in the literature, a clear delimitation is proposed to simplify and improve understanding and consistency:

Algorithm | Algorithm refers to the sequence of operations which must be executed to compute a response to a question (Merriam-Webster Dictionary). Algorithms are therefore tailored to the individual application and define a step-by-step procedure.

Method | Method is defined as *"a particular way of doing something"* (Cambridge Dictionary). Given this definition, a method represents the general principle and procedure for solving a mathematical problem. Methods have to be adjusted to and trained for specific applications via, e.g., (hyper-)parameter definition. *Technique* can be used as a synonym for method (Merriam-Webster Dictionary).

Approach | Lastly, an approach is defined as *"a way of considering or doing something"* (Cambridge Dictionary). Since this definition is not clearly distinguishable from the term method but resides on a slightly higher level of abstraction, in the following, the term approach depicts a thought process on how to generally solve a problem.

2.3.1 Prognostic Approaches

For the development of a prognostic algorithm, a broad spectrum of different methods is applicable (Coble and Hines Wesley 2011, p. 2). Selecting a method, however, depends strongly on the concrete application (Sikorska et al. 2011, p. 1803). While there are several ways to categorize existing methods (Elattar et al. 2016, p. 132), the most common classification assigns methods based on the information on which the model is built. The three common prognostic approaches are physics-based, data-driven, and hybrid (Lee et al. 2014b, p. 317; Guillén et al.

2016a, p. 994). Some more approaches, such as knowledge-based (Sikorska et al. 2011, p. 1809) or reliability-based (Elattar et al. 2016, pp. 133 f.), are presented by some authors. As methods within these approaches are either rarely addressed or can be subsumed under one of the three main approaches, they are not discussed further.

Physics-based approaches describe machine degradation by refining a set of algebraic and differential equations (Kan et al. 2015, p. 2; Elattar et al. 2016, p. 134). Thus, a thorough physical understanding of the machine, possible failure modes, as well as effects of influencing factors (e.g., operating conditions and load) is required, which has to be translated into a mathematical model (Heng et al. 2009, p. 726). Besides mechanical knowledge, the model is generally built on simulated data under nominal and degraded conditions (Lee et al. 2014b, p. 317). Physics-based approaches are commonly also referred to as model-based approaches (e.g., Kan et al. 2015, p. 2; Lee et al. 2014b, p. 317). However, this terminology is misleading as data-driven approaches entail the development of models as well. If the degradation behavior is well understood, physics-based models can outperform data-driven models (Sikorska et al. 2011, p. 1831) and might therefore be the preferred choice for applications in which accuracy is of utmost importance (Heng et al. 2009, p. 728). Nevertheless, physical models for industrial applications are difficult or even impossible to realize due to the complexity of real systems (Jardine et al. 2006, p. 1494). This is also attributed to the fact that physics-based models portray unique solutions for machines and fault types, which are reusable only to a very limited extent (Elattar et al. 2016, p. 134).

Data-driven approaches analyze historical condition monitoring data to identify the current machine condition and predict future trends (Kan et al. 2015, p. 3). Thus, data-driven methods explore degradation trends and adapt them to the current situation with little or no domain expertise (Heng et al. 2009, p. 728). For this purpose, data-driven approaches require (multivariate) data from multiple run-to-failure events, which represent different degradation patterns from the normal to the faulty state until system failure (Sun et al. 2012, p. 330). Due to the approaches' ability to derive hidden structures from data, they can model highly complex systems (Lam et al. 2014, p. 624; Zaidan 2014, pp. 29 f.). For industrial applications, data-driven approaches are often the preferred option as the inherent physics-of-failure of the system degradation in real-world systems is not extensively understood or difficult and expensive to gather (Hu et al. 2012, p. 121; Elattar et al. 2016). Data-driven methods are generally associated with low costs and fast realization (Elattar et al. 2016, pp. 134 f.). However, they come with higher computational requirements and perform poorly in cases where data is not representative (Kan et al. 2015, p. 3; Elattar et al. 2016, p. 135).

The *hybrid approach* combines two methods in order to integrate their strengths as well as overcome individual limitations (Atamuradov et al. 2017, p. 14). This can be done by fusing data-driven and physics-based methods or multiple data-driven methods (Atamuradov et al. 2017, pp. 14 f.). The fusion can be performed either pre-estimate (merge methods to generate a joint estimation) or post-estimate (combining results from different methods) (Elattar et al. 2016, p. 137).

The determination of a reasonable method depends on various criteria, such as the type and quality of the data, as well as assumptions that can be drawn about the degradation behavior (Coble and Hines Wesley 2011, p. 2). Thus, selecting a suitable prognostic algorithm requires a good understanding of the challenges associated with a specific application (Sikorska et al. 2011, p. 1804). A study on prognostic approaches by Lei et al. (2018, pp. 814 f.) revealed that most research had been done in the field of data-driven approaches (82%), while physics-based (10%) and hybrid approaches (8%) have only been addressed by the minority of publications. Due to the many advantages of data-driven approaches, in particular for practical implementations and their high generalizability, data-driven approaches are the focus of this thesis.

2.3.2 Data-Driven Methods

For the development of a data-driven prognostic method, two main approaches can be distinguished: direct RUL mapping and degradation-based mapping. *Direct RUL mapping* depicts a simple and efficient approach based on supervised learning (Jia et al. 2018, p. 3; Mosallam et al. 2015, p. 2). In this approach, a direct mapping is learned between a feature vector and an output label (usually RUL) (Ramasso and Saxena 2014, p. 8). A prevalent method representing this approach is an artificial neural network (Jia et al. 2018, p. 4). *Degradation-based mapping*, on the other hand, is a two-step procedure in which the degradation trend is initially determined (Huang et al. 2017, p. 14). Based on this knowledge, a prediction model is created by extrapolation up to a predefined alarm or failure threshold (Jia et al. 2018, p. 6; Khelif et al. 2017, p. 2277). The degradation trend can either be represented by one or multiple dominant physical signals (referred to as a physics health indicator) or in the form of a synthesized health indicator (Wang et al. 2012, p. 624). In particular, regression and curve-fitting methods, as well as state-space methods, are associated with this approach (Jia et al. 2018, p. 6). In contrast to direct RUL mapping, which implicitly assumes a linear degradation trend, it can take arbitrary shape in the degradation-based mapping (Jia et al. 2018, p. 6). The basic procedure of both approaches is outlined in Figure 16.

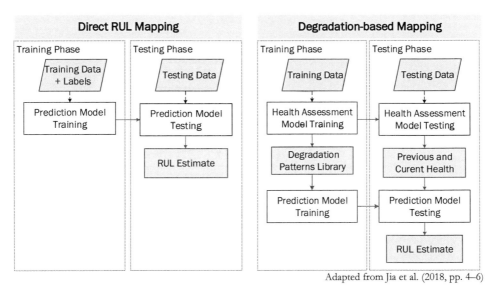

Adapted from Jia et al. (2018, pp. 4–6)

Figure 16 | General Data-Driven Prognostic Approaches

There exist numerous methods which can be used to build an application-specific prognostic model.

Table 1 provides an overview of common prognostic methods as presented by several review papers.[3] From these reviews, seven common prognostic methods are identified, which are at least mentioned by two reviews. The proportional hazard model was excluded despite being referenced three times as it represents a type-II prognostic method (Coble 2010, p. 16). The seven methods can be classified into stochastic, statistical, and machine learning methods. *Stochastic methods* represent stochastic processes that model the machine deterioration by a collection of random variables indexed by time (Sikorska et al. 2011, pp. 1816 f.). *Statistical methods* focus on fitting statistical models to historical data and infer relationships between variables (Bzdok et al. 2018, p. 233). Lastly, *machine learning methods* aim to find generalizable predictive patterns (Bzdok et al. 2018, p. 233). In contrast to statistical models, machine learning methods require minimal assumptions about the system and are better suited for complex, high-dimensional problems (Bzdok et al. 2018, p. 233). The presented classes are not mutually exclusive; however, they portray a good structure to highlight key characteristics.

---

[3] For the identification of established prognostic review publications, the ten most cited papers in the Scopus literature database were analyzed with respect to the title search query "review" AND ("prognostic" OR "remaining useful life"), excluding the field of medicine. In addition, the results are extended to include known prognostic reviews as outlined by ATAMURADOV ET AL. (2017, pp. 2 f.).

Table 1 | Overview of Data-Driven Methods in Review Papers

Review Publications	Stochastic		Statistical		Machine Learning			Additional Methods
	Stochastic Processes	Bayesian Networks/ Stochastic Filters	Trend Regression	Autoregressive Models	Artificial Neural Networks	Support Vector Machines	Gaussian Process Regression	
Atamuradov et al. (2017)		X		X	X		X	Gaussian mixture
Heng et al. (2009)		X	X		X			
Kan et al. (2015)		X		X	X	X	X	
Lee et al. (2014b)		X		X	X	X		Decision Tree
Lei et al. (2018)	X	X	X	X	X	X	X	*Proportional Hazard Model*
Peng et al. (2010)		X		X	X			*Proportional Hazard Model*, Grey Model
Rezvanizaniani et al. (2014)					X	X		
Si et al. (2011)	X	X	X					
Sikorska et al. (2011)		X	X	X	X			Reliability Functions, *Proportional Hazard Model*
Zhang and Lee (2011)		X				X		

For the selection of an appropriate method, it is necessary to examine the no-free-lunch theorem, which originates from the field of machine learning (Alpaydin 2019, p. 567). This theorem states that any machine learning algorithm has the same accuracy rate when classifying unlabeled data points across all data-generating distributions (Goodfellow et al. 2016, p. 114). For this reason, there is no generally preferred method, but it should be selected according to the underlying data distribution (Goodfellow et al. 2016, p. 116). The no-free-lunch theorem was adapted to prognostics by asserting that the superior performance of a method on a data set can be viewed as a better fit to the particular time-series data attributes; hence there is no universally superior prognostic method (Goebel et al. 2017, p. 66). The selection of a suitable method for a specific application is, thus, often not trivial and depends on the data and application characteristics (Sikorska et al. 2011, p. 1804). This step, however, is critical to the successful implementation of prognostics and can result, in the worst case, in project failings (Sikorska et al. 2011, p. 1803).

The seven methods are shortly presented in the following chapters with regard to the identified method classes, especially highlighting their key characteristics, main benefits, and drawbacks.

2.3.3 Stochastic Methods

The first class, stochastic methods, defines degradation as a stochastic process based on random variables. These methods are able to account for unexplained randomness of degradation behavior, which may be caused by unobserved or unknown influencing effects, such as intrinsic randomness and environmental factors (Ye and Xie 2015, p. 23). Stochastic methods are therefore well suited to characterize the uncertainties and can reflect the dynamics of the degradation progression (Lei et al. 2018, p. 816; Si et al. 2011, p. 6). There are two established stochastic methods in prognostics, namely simple stochastic processes and Markov models.

Simple Stochastic Processes model the degradation as a sequence of stationary independent increments over time depicted by a random variable (Xia et al. 2018, p. 260). Stochastic processes are therefore well suited to study the temporal variability of a degradation path (Lei et al. 2018, p. 816). Failure for prognostics is defined as the first time a predefined failure threshold is passed (Ye and Xie 2015, p. 16). Three types of stochastic processes are mainly considered in prognostics, namely the Wiener process with drift, the Gamma process, and the Inverse Gaussian process. A Wiener process with drift (often only denoted as Wiener process) is described by the weighted sum of a drift (linear trend) and a diffusion term following Brownian motion (Mishura and Shevchenko 2017, p. 36). Increments for the Wiener process are, therefore, normally distributed, which results in non-monotonic degradation behavior (Ye and Xie 2015, pp. 17 f.). Gamma processes and Inverse Gaussian processes, on the other hand, are strictly monotonous stochastic processes with only positive increments following the Gamma distribution, respectively the Inverse Gaussian distribution (Tsui et al. 2015, p. 6; Lei et al. 2018, pp. 816 f.). They are, therefore, well suited to describe wear and cumulative damage (Ye and Xie 2015, p. 20). Besides the basic formulation of these processes, there are three variants used in prognostics. This includes the consideration of measurement errors in the form of a time-independent error term, the implementation of random effects that model unknown population heterogeneity, and the analysis of environmental stress factors using covariates (Ye and Xie 2015, pp. 18–21).

The calculation of stochastic processes is relatively simple, and the physical interpretation is evident (Si et al. 2011, p. 6). Moreover, the failure distribution can be derived analytically (Si et al. 2011, p. 4). However, stochastic processes disregard historical observations and are memoryless (Lei et al. 2018, pp. 816 f.). In addition, Wiener processes are time-homogeneous, which is not always given in real applications (Si et al. 2011, p. 5).

Another set of methods are dynamic Bayesian networks and, in this context, in particular, *Markov models*. Unlike the above presented stochastic processes, these methods model the degradation using a finite number of states and related transition probabilities (Tsui et al. 2015, p. 6). States depict distinct levels of degradation from healthy to fully degraded, which are labeled either as "available" or "failed" (Sikorska et al. 2011, p. 1821; Tsui et al. 2015, p. 6). For time-series modeling, states are connected by directed arcs, which describe the time flow (Ghahramani 2001, p. 5). The order of a Markov model determines the number of previous observations, which are included in the transition probability (Ghahramani 2001, p. 2). A first-order Markov model (Markov chain) therefore expects that future behavior is only determined by the current state (Kan et al. 2015, p. 5). Like stochastic processes, Markov models are thus memoryless (Awad and Khanna 2015, p. 85). The basic Markov model assumes that the sojourn time (time spent in a state) of each state is exponentially distributed, and changes are time-invariant (Xia et al. 2018, p. 260). Semi-Markov models depict a generalization in which the sojourn time follows any distribution (Sikorska et al. 2011, p. 1821). While Markov models can describe direct condition data, a hidden Markov model (HMM) is considered in the case of indirect data (Si et al. 2011, p. 3). As the states are not directly observable in this case, the degradation is described by a doubly embedded stochastic model. To each hidden state, a set of observed measurements (observed states) are thus assigned. The unobservable hidden state sequence can be inferred by the sequence of observations (Awad and Khanna 2015, p. 84). Analogous to the basic model, HMMs can be generalized to semi-HMMs, which allow any distribution of the sojourn time for the unobserved state model (Si et al. 2011, p. 10). To use Markov models for prognostics (RUL estimation), the sojourn times of future states until a failure state is reached have to be summed (Anger 2018, p. 62).

Stochastic filters, in particular, Kalman and Particle filters, are closely related methods. While in HMMs, the hidden state variable is discrete, stochastic filters are used in continuous state-space models where the hidden variables are replaced by real-valued variables (Si et al. 2011, pp. 9 f.; Ghahramani 2001, p. 6). They enable state estimation and are, therefore, also used in physics-based approaches (Sikorska et al. 2011, pp. 1822 f.).

Markov models are easy to interpret as existing states have a clear meaning (Sikorska et al. 2011). In addition, they can be exclusively learned with data or based on physical knowledge (Sikorska et al. 2011, pp. 1819–1821). For data-driven development, incomplete data can be exploited (Xia et al. 2018, p. 260). Compared to that, the applicability of the Markov models is limited by the underlying assumptions. This includes the Markov property as well as the time-invariance (Kan et al. 2015, p. 6). The latter can be mitigated by the use of semi-Markov models (Peng et al. 2010, p. 305). Furthermore, training requires a large amount of data to learn transition probabilities between states (Si et al. 2011, p. 7).

2.3.4 Statistical Methods

Statistical methods belong to the field of time-series analysis. In contrast to the above described stochastic methods that focus on the determination of the structure of the random variables, statistical methods and machine learning aim at regression. Statistical methods develop a mathematical model based on inference (Bzdok et al. 2018, p. 233). They can include system knowledge and can verify assumptions. Statistical methods are, therefore, particularly suited for small to medium sample sizes with low dimensionality (Bzdok et al. 2018, p. 233). Simple trend regression and autoregressive models are representative methods of this class in prognostics.

Simple Trend Regression is considered one of the simplest and most popular methods in the industry (Sikorska et al. 2011, p. 1823; Si et al. 2011, p. 3). This method takes as input a single condition monitoring variable, which depicts the current state of degradation (Sikorska et al. 2011, p. 1823). The variable represents either a key condition monitoring sensor or aggregated values with regard to different sensors. Taking into account the degradation behavior, a reasonable data representation model is selected and fitted to the data by estimating related parameters (Adhikari and Agrawal 2013, p. 15). For prognostics, the remaining useful life depicts the time until the variables exceed a predefined failure threshold (Si et al. 2011, p. 3). This can be realized by inter- or extrapolation of the fitted data representation model (Sikorska et al. 2011, p. 1823).

There are various data representation models that can be used for prognostic applications, in particular linear and exponential models (Si et al. 2011, p. 3; Heng et al. 2009, p. 728). Besides these simple models, another variation covers random coefficient models (also denoted as a general path model or mixed-effects models). Instead of assuming deterministic failure patterns among all degradation paths, these models include random effects with regard to, e.g., different usage patterns or environmental influences (Tsui et al. 2015, p. 7; Ye and Xie 2015, p. 22). To capture the random effects of specific degradation behavior, random coefficient models can be extended with Bayesian inference (Coble and Hines Wesley 2011, pp. 74 f.).

Due to its simplicity and low data requirements, the method simple trend regression has high practical relevance (Sikorska et al. 2011, p. 1824; Lee et al. 2017, p. 17). However, they require a clearly defined degradation trend with no or little influence by usage or environment (Sikorska et al. 2011, p. 1815). For this purpose, a simple trend regression often oversimplifies the failure trajectory. In particular, deviations from monotonic behavior cannot be captured adequately (Heng et al. 2009, p. 728). Their performance, therefore, highly depends on the fitness of the selected model to the data (Lee et al. 2017, p. 17).

A method similar to simple trend regression denotes the family of *Autoregressive Models*, which includes the autoregressive moving average (ARMA) model and its variations. These models make predictions using a linear function that accounts for observations and random errors (Lei et al. 2018, p. 816). For this purpose, the ARMA model linearly combines an autoregressive component, a moving average component, and a random error. The autoregressive component yields the weighted moving average of the p past observations, whereas the moving average component calculates the weighted moving average of the q past errors (Box et al. 2016, pp. 75–77). The order (p and q) of both components thus denotes the number of historical values, which are considered for prediction. These can be estimated using the (partial) autocorrelation function (Box et al. 2016, pp. 77 f.). To develop an ARMA model, first, the order of the auto-regressive and moving average components are defined (model identification), followed by parameter estimation (weights) using non-linear optimization techniques. At last, model validation is performed on unseen data (Sikorska et al. 2011, p. 1824). The application of ARMA for the RUL estimation is analogous to the procedure explained for the simple trend regression by predicting future values until a failure threshold is reached.

The application of ARMA models requires stationarity of the data. Stationarity is given when the joint distribution of any set of observations is not affected by a time shift (Box et al. 2016, p. 24). In prognostics, this assumption rarely applies. The variant ARIMA (autoregressive integrated moving average) model, therefore, presents an adequate model to address trends in time-series through differencing (Sikorska et al. 2011, p. 1824). Other common extensions in prognostics are the seasonal autoregressive integrated moving average (SARIMA) model, which features a seasonal component for waveform data, the autoregressive moving average model with exogenous inputs (ARMAX) model, which incorporates external factors, as well as the vector autoregressive moving average (VARMA) model, which handles multivariate data (Box et al. 2016; Sikorska et al. 2011, p. 1824).

Autoregressive models are very well suited for short-term prediction (Xia et al. 2018, p. 259). They feature computational efficiency and require no prior knowledge on the failure trajectory (Lei et al. 2018, p. 816; Sikorska et al. 2011, p. 1813). In contrast, autoregressive models show bad performance when making long-term predictions due to their sensitivity to noise and initial conditions, as well as the systematic error accumulation (Box et al. 2016; Sikorska et al. 2011, p. 1813). Furthermore, the assumption of linearity and stationarity for the ARMA base model poses the key limitation for its applicability in prognostics (Xia et al. 2018, p. 260; Lee et al. 2014b, p. 323).

2.3.5 Machine Learning Methods

The term machine learning was coined by Samuel (1959). In this work, he described machine learning as "*[p]rogramming computers to learn from experience [to] eventually eliminate the need for much of this detailed programming effort*" (Samuel 1959, p. 211). While some authors classify machine learning techniques as statistical methods (e.g., Breiman 2001), there are fundamental differences that justify an additional mentioning. In general, statistics, as described above, focus on inference from a representative population, whereas machine learning aims to detect patterns from data (Bzdok et al. 2018, p. 233). In cases where there is limited knowledge about the system and its degradation, and the available machine data is high-dimensional, unwieldy, and complex, machine learning becomes the preferred choice (Bzdok et al. 2018, p. 233; Bektas et al. 2019, p. 88). However, machine learning lacks an explicit model (so-called black box) and thus does not provide information on the underlying behavior to draw further conclusions (Bzdok et al. 2018,

p. 233). Machine learning methods are typically divided into supervised, unsupervised, and semi-supervised, depending on the availability of labels (Sun et al. 2012, p. 330). Common representative machine learning methods for prognostics are artificial neural networks (and their variants), vector machines (including support and relevance vector machines) as well as gaussian process regression.

Artificial Neural Networks (ANN) are the most frequently chosen method for data-driven prognostics (Heng et al. 2009, p. 729; Lei et al. 2018, p. 818). Their structure is inspired by the human brain (Lee et al. 2014b, p. 323). The network is composed of three types of layers (input, hidden, and output layers), which consist of one or multiple processing elements (nodes or neurons) (Bishop 2009, p. 228). A node receives (weighted) inputs, modifies them using an activation function, and transmits the results to the succeeding node (Goodfellow et al. 2016, pp. 164 f.). The network weights are trained using learning algorithms, such as backpropagation (Bishop 2009, p. 241). ANNs can be applied in different ways for prognostics; this includes time-series step prediction and direct RUL prediction (Kan et al. 2015, p. 8). The former refers to a single-step or multiple-step prediction of a health indicator (regression), which is extrapolated until a pre-defined threshold is reached, while the latter depicts the direct failure time prediction (Heng et al. 2009, p. 729; Sikorska et al. 2011, p. 1830). In addition, ANNs can also be used as a supporting method to estimate the parameters of a known function (Sikorska et al. 2011, p. 1830).

There exist various ANN network structures, which can generally be divided into static and dynamic architectures (Kan et al. 2015, p. 8; Sikorska et al. 2011, p. 1827). Static networks are also referred to as feed-forward networks, which are characterized by node connections between adjacent layers without allowing for cycles (Sikorska et al. 2011, p. 1827). Feed-forward neural networks are typically developed to learn a direct mapping between the current health condition and the RUL estimate (Lei et al. 2018, p. 818). Well-known architectures are the multi-layer perceptron as well as the radial basis function network (Kan et al. 2015, p. 8; Sikorska et al. 2011, p. 1827). While static networks are memoryless, information can be stored in dynamic networks and accessed in subsequent iterations. Dynamic networks are, therefore, capable of explicitly analyzing time-series data and learning temporal structures (Lei et al. 2018, p. 818). These networks are called recurrent neural networks. Here, long short-term memory and time-delay neural networks are popular variants. In particular, recurrent neural networks are eligible in single- and multiple-step predictions of the health indicator (Kan et al. 2015, p. 8). Besides feed-forward and recurrent neural networks, deep neural networks are frequently applied. Deep neural networks do not imply a different network structure but rather refer to the existence of multiple hidden layers (Goodfellow et al. 2016, p. 163). The depth of the network is leveraged for the automatic learning of the data representation, replacing manual feature extraction (LeCun et al. 2015, p. 436).

2.4   Current Research Gaps in Predictive Maintenance

To conclude, four research gaps are derived from this chapter, which will be addressed in more detail in the following research of this thesis.

Research Gap 1:

Lack of detailed support in the realization of predictive maintenance projects

A considerable amount of research has been dedicated to the development of PdM solutions. Most of these works, however, are explorative and, to a large extent, application-specific (Saxena et al. 2010a, p. 2; Cocheteux et al. 2009, p. 2). Furthermore, the data analysis process, as introduced in Chapter 2.2.1 and detailed in the following chapters, is almost exclusively emphasized. While there are already some generalized cross-application standardization efforts in this direction, the realization of predictive maintenance projects in their entirety is hardly considered at present. This research gap is addressed by the overall research objective of this thesis. In particular, a process reference model for the implementation of a predictive maintenance project is developed in Chapter 3, which systematically guides users in the development of an application-specific process model. This is extended in Chapters 4 and 5 with a generic method for fleet prognostics (type-IV prognostics).

Research Gap 2:

Lack of understanding of the use of fleet data in predictive maintenance,
and in particular in prognostics

Predictive maintenance, and in particular data-driven diagnostics and prognostics, depends to a large extent on the availability of data. Therefore, as outlined in Chapter 2.2.2, the use of fleet data for predictive maintenance bears the significant potential to improve the quality of diagnostic and prognostic algorithms in cases with low data representativeness. However, initial research on this topic has so far been very fragmented and focused on specific applications. Thus, in order to leverage fleet data for predictive maintenance, in the first place, the concept of fleets within predictive maintenance and the key challenges involved must be better understood. This concerns, in particular, the possible heterogeneity of the machines within a fleet. This research gap is addressed by Chapter 4 through a conceptual analysis of general fleet characteristics and an elaboration of the associated challenges in handling fleet data for predictive maintenance, along with a proposal for characterizing them from a data-driven perspective.

Research Gap 3:

Lack of knowledge on processing homogeneous and heterogeneous data from multiple
machines for the purpose of prognostics

Building on the previous research gap, the processing of data from multiple machines (fleets) poses a further challenge due to the heterogeneous nature of machines originating from various influencing factors, such as customer-specific machine configurations (Voisin et al. 2015, p. 138) and/ or different operating and environmental conditions (Al-Dahidi et al. 2017, p. 2). This represents the type-IV prognostics, as introduced in Chapter 2.2.5.1. Even though several approaches exist that leverage fleet data for prognostics (Tsui et al. 2015, p. 7), most methods are designed for specific fleet compositions and application characteristics. As of yet, however, it is unclear which method should be used for specific fleet characteristics and how it should be implemented. To support the implementation of fleet prognostics (type-IV prognostics), a generic method for the development of prognostic models taking into account fleet data is targeted in Chapter 5, which portrays different fleet prognostic approaches and guides in a step-by-step manner through their implementation.

Research Gap 4:
Lack of guidance regarding the selection of methods, taking into account
data and application characteristics

As described in the previous subsection (Chapter 2.3), there are a variety of different methods for computing the remaining useful life of machines. While there is no general superiority of one method (cf. Chapter 2.3.2), each of them features multiple advantages and disadvantages. A few of them have been listed above. Depending on the application and the particular data, the methods are therefore more or less suitable to meet the specific application requirements. However, to date, the choice of methods for the development of a prognosis model for a concrete application is often arbitrary and not justified in detail (Lee et al. 2014b, p. 319). To enable a more informed decision, the research gap of missing guidance for the selection of a prognostic method for a given application is targeted by a multi-criteria generic method selection process in Chapter 5. Even though the process mainly aims at method selection for prognostics with multiple machines (fleets), some results can also be exploited for prognostics in general.

tion of a context-specific model (Rosemann and van der Aalst 2007, p. 2) and can yield improved business processes at decreased cost and risk (Matook and Indulska 2009, p. 60). In this context, a process reference model is denoted as a specific type of reference model (Fettke et al. 2006, p. 469).

Process models depict formal representations of processes (Hickey and Davis 2003, p. 1). With regard to businesses, processes are defined as the *"completely closed, timely and logical sequence of activities which are required to work on a process-oriented business object"* (Becker et al. 2011, p. 4). They have a clear start and end point with pre-defined in- and outputs (Davenport 1993, p. 5). Based on this definition, business processes are understood to be specific processes that are driven by a business objective and the organizational and technical business environment (Becker et al. 2011, p. 4; Weske 2012, p. 4). A reference process addresses several aspects mentioned for reference models, however, with regard to a process instead of a model. They are defined to formalize recommendations for a complete set of processes in a specific domain and are in this regard customizable for a specific context (Hallerbach et al. 2008, p. 7; Schütte 1998, pp. 69 f.; Santana et al. 2014, p. 249).

In this work, the term *process reference model* is used in view of a process model with reference character and thus complies with the two definitions presented above. The developed process reference model describes a comprehensive process model for the implementation of predictive maintenance. It presents a set of steps, aligned as a process that covers a wide range of different predictive maintenance implementation processes. For a specific application, a reduced and adapted process can be deduced from the process reference model. In order to develop the process reference model for the implementation of a predictive maintenance solution, defined characteristics for process reference models must be adhered to. By studying available literature, a comprehensive set of derived characteristics is presented by Matook and Indulska (2009), which aim to provide general guidance for the development of process reference models.

Generality: Process reference models must have a certain degree of generality. They require an abstract perspective in order to be applicable in different settings. To ensure general validity, a process reference model is defined for a specific scope and therefore developed for cases with comparable process characteristics (Matook and Indulska 2009, pp. 62 f.).

Flexibility: Process reference models must be flexibly adaptable to different application characteristics (Matook and Indulska 2009, p. 63). Adaptations and modifications for individual organizational needs are necessary. To allow flexibility, a modular design of the process reference model is desirable (Schwegmann 1999, p. 71).

Completeness: Process reference models must be complete and correct. The essential process elements have to be covered, and the different application characteristics are handled. It should, therefore, adhere to the users' requirements (Moody and Shanks 1994, p. 98; Matook and Indulska 2009, p. 63). The process reference model provides relevant knowledge if at least one application case is conceivable (Scheer 1999, p. 8).

Usability: Process reference models should exhibit a high degree of usability. While meta-models provide guidance on an abstract level, process reference models must exhibit a low degree of granularity to enable their implementation (Hars 1994, pp. 15 f.; Moody and Shanks 1994, p. 106; Matook and Indulska 2009, p. 63).

Understandability: Process reference models must be simple to understand by the model user and structured in a comprehensible way with regard to the specifics of the target group (Matook and Indulska 2009, p. 62). To enable their reusability, process reference models require a high acceptance by the user (Schwegmann 1999, p. 62).

3.2 Methodological Approach

The development of the process reference model for the implementation of predictive maintenance is constructed following the *seven-step reference model development process* by Matook and Indulska (2009, pp. 61 f.). While there are several procedures for developing reference models, each with its own focus, this procedure synthesizes the existing established procedures and represents a coordinated view of process reference model development that is internationally known and widely cited in the research community. The methodological approach used for the development of the process reference model is depicted in Figure 17. The seven steps are as specified in the following:

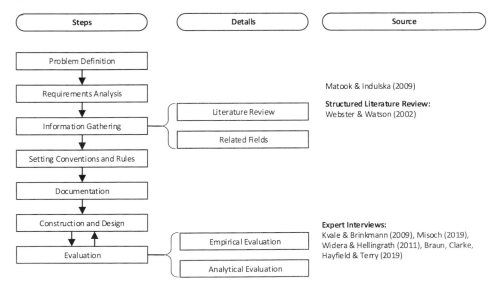

Figure 17 | Methodological Approach (RQ1) based on Matook and Indulska (2009)

Setting Conventions and Rules | The fourth step defines the conventions and rules to be followed during the design of the process reference model. This includes both the selection of a suitable modeling technique as well as the definition of the standardized syntax.

The selection of the modeling language should be guided by the objective of the model (Fleischmann et al. 2018, pp. 71 f.). For this purpose, the selection is driven by the two modeling requirements, as specified above. The four established process modeling techniques, namely flow charts, extended event-driven process chains, unified modeling language (UML) activity diagram, as well as business process modeling notation (BPMN) (Fleischmann et al. 2018, pp. 72 f.), are analyzed comparatively with respect to the two defined modeling requirements. Table 2 highlights the results of the assessment. With regard to the first modeling requirement, it can be seen that all languages, except the UML activity diagram, are able to model the required elements. For flow charts, the data/ document symbol, however, is only available as part of the additional flow chart data symbols, defined by the ISO 5807. In addition, for BPMN, the reference elements only depict references to sub-processes. The second modeling requirement, which targets the ease of understanding is best satisfied by the modeling language flow charts. Both within complexity and training needs, it shows that flow charts are easy to build and simple to handle. They can be understood without prior knowledge or training, whereas this is not the case for the other three languages. This is due to the fact that the other languages aim at IT or specialist departments (Gadatsch 2017, p. 102). The process reference model, however, should be of value to a larger audience. Since the flow charts modelling language fulfills both defined modelling requirements, this language is selected for the process reference model.

Table 2 | Comparison of Modeling Languages
(based on Fleischmann et al. 2018, pp. 125 f.; Gadatsch 2017, pp. 100–103; Lobe, pp. 111 f.)

Modeling Language	Start/ End Element	Process Element	Data/ Document Element	Reference Element	Decision Element	Flow Element	Distribution	Complexity	Training Needs
Flow Charts	✓	✓	(✓)	✓	✓	✓	International	Low	Low
Extended Event-Driven Process Chain	✓	✓	✓	✓	✓	✓	DACH countries	High	High
UML Activity Diagram	✓	✓	✗	✗	✓	✓	International	High	Medium
BPMN	✓	✓	✓	(✓)	✓	✓	International	Very high	Very high

Regarding the syntax of the naming of the used elements, the following uniform rules are defined as depicted in Table 3:

Table 3 | Modeling Element Symbols and Syntax

Modeling Element	Symbol	Syntax	Example
Process Element	Process Element	[Noun]	Candidate Identification
Start/ End Element	Start/ End Element	[Begin | End] Process Name	Begin Preparation Phase
Decision Element (Iteration)		For each [object]:	For each candidate:
Decision Element (Check)	Decision	[Subject to be checked] [Adjective] ?	Performance acceptable?
Decision Element (Selection)		Select	Select
Reference Element (First Level)		[Process]	Candidate Solution Deployment
Reference Element (Second level)	Reference Element	[Process]: [Process]	Project Preparation: Candidate Identification
Data/ Document Element	Data/ Document Element	[Noun]	Project Plan
Process Flow	⟶		
Information Flow	⇢		

Documentation | The fifth step denotes the documentation of the key decisions. In this research, this is primarily done by the considered models originating from the literature as well as from related fields, which were used during the construction process. Furthermore, the adjustments resulting from the evaluation loop are documented and highlighted to ensure that the design is transparent and reproducible.

Construction and Design | The construction of the process reference model, depicted by the sixth step, is founded on the knowledge base identified within the information-gathering step. For this purpose, the existing knowledge was synthesized and arranged in a hierarchical structure to account for different levels of abstraction of the overall process. These levels of abstraction range from the most general to the most detailed and are referred to as phases (level 1), processes (level 2), and process elements (level 3).[6] In an iterative process, the identified process models from the literature are incorporated into the process reference model and aligned, as depicted in Figure 18. In the first step, the high-level synthesis, the identified process models which target the complete PdM realization process (classified by the general scope attribute) are

---

[6] The hierarchical structure is detailed in Chapter 3.4 as part of the introduction of the developed process reference model.

synthesized. This results in a definition of key phases (level 1) and major processes (on level 2). Thereafter, in a second step, the process models that refer to a single part of the overall model are successively integrated with respect to two further levels of detail, namely processes (level 2) and process elements (level 3). In the event of discrepancies at a higher level, the prior defined phases and processes were refined. The resulting three-level process reference model was matched against application cases in the third step. The resulting deviations were integrated unless they were deemed case-specific. In the final step, the process reference model was iteratively extended by drawing on existing methodologies from related fields.

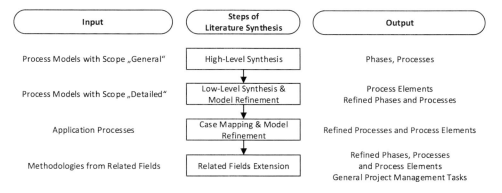

Figure 18 | Literature Synthesis Process

Evaluation | The final step comprises the evaluation of the process reference model in terms of its value and usefulness (Fettke and Loos 2003, p. 81). For this purpose, both an analytical and an empirical evaluation of the developed process reference model are conducted. Following the development-evaluation loop as specified by the design science approach in Chapter 1.2, the potential need for adjustment resulting from this evaluation is used to further improve the quality of the process reference model.[7]

3.3 Information Gathering

The following sub-chapter presents the identified knowledge base on which the process reference model is constructed. For this purpose, first, the models identified in the literature review are outlined (Chapter 3.3.1). Subsequently, selected methodologies from related fields are presented, which further contribute to enhancing the available knowledge base (Chapter 3.3.2).

---

[7] The final resulting process reference model, which incorporates the minor adjustments identified during the evaluation, is provided in the Appendix C and D.

3.3.1 State of the Art in Research

Existing process models from the literature are hardly or not applied since current implementations of predictive maintenance in the industry are carried out either ad-hoc or based on their own developed processes. Nevertheless, there are several models described in academic research as well as in standard documents, which describe (parts of) the predictive maintenance implementation process.

Predictive maintenance processes outline the sequence of steps to be performed to realize an implementation. While PdM process models depict an abstraction to illustrate the generic process sequence, processes refer to specific instantiations. Processes and process models identified within the literature can be grouped into four groups and two levels of detail, depending on their origin and focus. Processes and process models are characterized as complete if they cover the entire PdM realization process, while those labeled detailed describe only one facet of the overall process, but usually at a greater level of detail. Figure 19 depicts a visualization of the different groups.

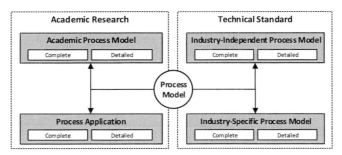

Figure 19 | Process Model Classification

The first category represents the *academic process models*. Academic research in this category develops conceptual process models to highlight and address aspects of the implementation process. On the general level, this category includes six application-independent process models: the 5s methodology, which describes five processes for PHM implementation (Lee et al. 2009b; Lee et al. 2014b; Lee et al. 2017), the generic PHM system development process (Vogl et al. 2014b, 2014a; Vogl et al. 2016), the systematic approach for predictive maintenance service design (Lee et al. 2009a), the Gantt diagram for a predictive maintenance program (Carnero Moya 2004), the design process for a prognostics solution (Goebel et al. 2017) as well as the procedure model for the introduction of predictive maintenance based on business intelligence (Bachmair 2018).

In addition, a large number of detailed process models are available, in particular with regard to PHM and prognostics in particular. Within the identified general PHM processes, different steps are highlighted, from data acquisition to decision making. While the general process is very similar, differences are in regards to the level of detail and design of the steps. Identified procedures are the PHM steps (Atamuradov et al. 2017), condition-based maintenance procedure (Shin and Jun 2015; Peng et al. 2010), the systematic maintenance decision framework (Peng et al. 2010), PHM functional architecture (Bonissone 2006), typical flows of PHM systems (Tsui et al. 2015), the PHM methodology developed by the Center for Advanced Life Cycle Engineering (CALCE) and the steps involved in general data-driven prognostic methodology (Pecht and Kumar 2008; Kumar et al. 2008) as well as different elaborations on the OSA-CBM (open standard architecture for condition-based maintenance) architecture standard proposal based on the ISO 13374 (Lebold et al. 2003; Bengtsson 2004; Funk and Jackson 2005; Zhang et al. 2015). Considering specific prognostic processes, these models include the conceptual framework for prognosis developed based on the integrated system of proactive maintenance (SIMP) (Voisin et al. 2010; Muller et al. 2008; Léger and Morel 2001; Bousdekis et al. 2015a, 2018), the ADEPS (Assisted Design for Engineering Prognostic Systems) methodology (Aizpurua and Catterson 2015, 2016), major components of prognostics implementation (Sun et al. 2012), the five generic process steps for a prognostic program (Li et al. 2018b) as well as the four technical processes in a machinery health prognostic program (Lei et al., 2018).

Furthermore, research has been conducted on various parts of the overall PdM process, with regard to the project preparation (Uckun et al. 2008; Johnson 2012; Guillén et al. 2016b), requirements specification (Saxena et al. 2010b), functional architecture definition (Li et al. 2018c), systems engineering (Millar 2007; Lamoureux et al. 2015), failure selection (Cocheteux et al. 2009), fault detection (Lapira 2012), diagnosis (Levrat et al. 2008), and decision making (Bousdekis et al. 2015a, 2018).

The second group depicts *process applications* presented in academic research. In these studies, a process is used to highlight the steps carried out for a specific case or for a developed PHM method. Although a large amount of literature is available within this group in the field of PHM, an exemplary subset was chosen to represent the group. Studies were selected which exhibit a detailed process of the steps carried out. These publications include several method-specific processes (Wang et al. 2012; Mosallam et al. 2015; Bonissone and Goebel 2002; Wang et al. 2008; Benkedjouh et al. 2013) as well as domain-specific processes (Bey-Temsamani et al. 2009; Das 2015; Shi et al. 2016).

The third and fourth group target existing standards. The former group comprises *industry-independent standards*, whereas the latter encompasses *industry-specific standards*. The fundamental standard document in predictive maintenance is ISO 17359. This standard provides a general procedure for condition monitoring. In addition, standards for diagnostics (ISO 13379) and prognostics (ISO 13381) are available. Furthermore, ISO 13374 explains the well-established open standard architecture for condition-based maintenance (OSA-CBM) in industrial applications

(Bousdekis et al. 2015b). For industry-specific standards, on the other hand, the Aeronautical Design Standard Handbook (ADS 79D-HDBK) from the U.S. Navy for aircraft as well as the IEEE Std 1856-2017 for electronic systems describe process models for their specific application areas.

From the set of identified processes and process models, it can be deduced that the majority of the models are either application-oriented or address a specific part of the overall process (in particular, targeting only the data analysis process). In addition, process models that depict the entire implementation process remain on a rather high level without providing detailed guidance. A detailed overview of the identified publications and standards is provided in terms of a concept matrix (cf. Table 4) with regard to their classification, the targeted scope, and focus, as well as their contribution to the elaborated phases of the PdM process reference process (namely project preparation, system architecture design, PHM design, candidate deployment phase, and project completion), which will be introduced in detail in Chapter 3.4.

Table 4 | SLR Concept Matrix

Concepts			General Characteristics				Covered Process Reference Model Phases				
			Scientific Publication	Standard	Scope	Focus	Project Preparation	System Architecture Design	PHM Design	Candidate Solution Deployment	Project Completion
Scientific Process Models	General	Bachmair (2018)	S		General		X		X		
		Carnero Moya (2004)	S		General		X	X	X		
		Goebel et al. (2017)	S		General	P	X		X		
		Lee et al. (2009a)	S		General		X		X		
		Lee et al. (2009b; 2014b; 2017)	S		General		X	X	X	(X)	
		Vogl et al. (2014b, 2014a; 2016)	S	(X)	General		X	X	X	X	
	Details	Guillén et al. (2016b)	S	(X)	Preparation		X	(X)			
		Johnson (2012)	S		Preparation	P	X		(X)		
		Uckun et al. (2008)	S		Preparation	P	X		(X)		
		Saxena et al. (2010b)	S		Requirements		X				
		Cocheteux et al. (2009)	S		Failure Selection	P	X		X		
		Li et al. (2018c)	S		Functional Architecture			X			
		Millar (2007)	S		Systems Engineering			X			
		Lamoureux et al. (2015)	S		Systems Engineering			X			
		Lapira (2012)	S		Fault Detection	FD			X		
		Levrat et al. (2008)	S		Diagnosis	D			X		
		Aizpurua and Catterson (2015, 2016)	S		Prognosis	P	(X)		X		

		Lei et al. (2018)	S		Prognosis	P				X	
		Li et al. (2018b)	S		Prognosis	P				X	
		Sun et al. (2012)	S		Prognosis	P	(X)			X	
		Atamuradov et al. (2017)	S		PHM					X	
		Bonissone (2006)	S		PHM					X	
		Lebold et al. (2003) Bengtsson (2004) Funk and Jackson (2005) Zhang et al. (2015)	S	(X)	PHM				(X)	(X)	
		Pecht and Kumar (2008) Kumar et al. (2008)	S		PHM					X	
		Peng et al. (2010)	S		PHM					X	
		Shin and Jun (2015)	S		PHM					X	
		Tsui et al. (2015)	S		PHM					X	
		Léger and Morel (2001) Muller et al. (2008) Voisin et al. (2010) Bousdekis et al. (2015a, 2018)	S		Prognosis/ PHM	P				X	
		Bousdekis et al. (2015a)	S		Decision Making	P				X	
Application Process	Details	Das (2015)	A		PHM					X	
		Wang et al. (2008)	M		Prognosis	P				X	
		Bonissone and Goebel (2002)	M		Prognosis	P				X	
		Bey-Temsamani et al. (2009)	M		Prognosis	P				X	
		Mosallam et al. (2015)	M		Prognosis	P	(X)			X	
		Wang et al. (2012)	M		Prognosis	P				X	
		Shi et al. (2016)	A		Diagnosis	D				X	
		Benkedjouh et al. (2013)	A		PHM	P	(X)			X	
Standards	General	ISO 17359	I		General	CM/D	X			X	
		ISO 13379	I		General	CM/D	X			X	
		IEEE Std 1856-2017	S		General		X	X		X	
		ADS 79D-HDBK	S		General	CM	X	(X)		X	
	Details	ISO 13381	I		Prognosis	P				X	
		ISO 13374	I		Data A. OSA-CBM			X			

Legend: Scientific Publications (S= Scientific PM, A= Application/ Case Study, M= Method), Standard (I= Industry-Independent Standard, S= Industry-Specific Standard, (X)= Scientific Publications referencing Standards), Focus (CM= Condition Monitoring, FD= Fault Detection, D= Diagnostics, P= Prognostics)

3.3.2 Methodologies in Related Fields

The literature and practical findings already demonstrated good coverage of the steps for the introduction of predictive maintenance. In some cases, however, the identified processes offer little detailed information on the concrete steps. Besides the integration of additional knowledge, the purpose is to align the elaborated process reference model with standard processes in related fields. The related fields include data analytics, business process improvement, systems engineering, and project management. The identified acknowledged processes within the related fields represent methodologies. These are defined as process models accompanied by a detailed description for each element, mainly focusing on the techniques and tools (Hickey and Davis 2003, p. 1). In order to distinguish them from the process models described above, they are therefore referred to in the following as methodologies.

Data Analytics | (in this context, often referred to as Knowledge Discovery in Data or Data Mining) aims at extracting information in large data sets (Fayyad et al. 1996, p. 82). Existing processes provide support for generic steps for a wide range of data analytics projects. Due to the need for sensor data analysis for diagnosis and prognosis, PdM implementation can be seen as a data analytics project and can, therefore, profit from existing knowledge. In the field of data analytics, the cross-industrial standard process for data mining (CRISP-DM) (Chapman et al. 2000) is chosen for integration. CRISP-DM is designed by an industrial consortium and depicts the leading methodology for data analytic projects in industrial applications (KDnuggets 2014; Mariscal et al. 2010, p. 138; Jahani et al. 2016, p. 209). It contains 24 process steps which are structured in six phases, ranging from the business perspective to the deployment.

Business Process Improvement | targets business process re-engineering in order to adapt and redesign existing business processes (Adesola and Baines 2005, pp. 38 f.). The implementation of predictive maintenance requires maintenance processes to be proactive. For this reason, described steps provided by the generic and practical methodology for model-based and integrated process improvement (MIPI) (Adesola and Baines 2005) are considered in the reference process model. The MIPI methodology was developed by reviewing and analyzing available methodologies and was thoroughly validated by industry interviews and case studies. It consists of a seven-step procedural approach that guides a process design team during business process improvement.

Systems Engineering | depicts an approach that targets the design and management of complex technical systems (Walden et al. 2015, p. 6). It includes the development of appropriate technology architecture, referred to as system architecture, which comprises the logical and physical view, as well as integration and operation. In order to design a PdM system architecture that fits best to the company's needs, systems engineering provides support from system concept development to system disposal (Walden et al. 2015, p. 2). The INCOSE (International Council on Systems Engineering) systems engineering handbook describes the key system engineering processes (Walden et al. 2015, p. 3). The handbook is compliant with the ISO/IEC/IEEE 15288:2015. It comprises 25 process steps within four process groups (enterprise, agreement, project, and technical processes).

Project Management | captures the different processes which are necessary to manage a project from its initiation to the closing. The implementation of predictive maintenance in a company represents a project as it creates a unique result and because its time duration is limited (Project Management Institute 2017, p. 4). As a result of the complexity and large scope of PdM projects, a good implementation of project management is essential for a successful realization. The Project Management Body of Knowledge (PMBOK) (Project Management Institute 2017) defines guidelines and standard terminologies developed by the Project Management Institute (PMI) and is reflected in ISO 21500:2012. The methodology provides well-established and widely-applied traditional practices (Project Management Institute 2017, pp. 1 f.). In the PMBOK, 47 process steps are elaborated, which are grouped into ten knowledge areas (from project integration management to project stakeholder management) and five process groups (initiation, planning, execution, monitoring, and controlling as well as closing).

3.4 Process Reference Model Design and Construction

This chapter presents the derived Process Reference Model for Predictive Maintenance (PReMMa).[8] PReMMa describes phases, processes, and process elements with their respective inputs and outputs for the realization of PdM implementations. The model is organized hierarchically on three levels of detail. The first level describes a phase model that includes five generic phases to be followed during the implementation of predictive maintenance. These phases are detailed on the second level, the process level. For each phase, a number of processes are highlighted. On the third level, the process element level, the identified processes are described with their corresponding steps. Figure 20 provides a visualization of the hierarchical structure of the process reference model.

PreMMa synthesizes literature from the field of predictive maintenance as well as from four related fields data analytics, business process improvement, systems engineering, and project management. Since project management is carried out continuously in parallel with a predictive maintenance project and the specific design depends largely on the intended project management approach, a synthesis in the hierarchical process structure was, in most parts, impossible. Instead, the five project management phases (initiation, planning, execution, monitoring and control, and closure) were aligned with the phases of a predictive maintenance project. For this reason, the project management phases are only assigned to the phases of the process reference model.[9] In contrast, the expertise from the other three related fields could be directly integrated into the model.

---

[8] The process reference model presented here reflects the model developed from the existing body of knowledge. Minor adjustments were identified during the empirical evaluation which will be described in Chapter 3.5.2.2. These were subsequently integrated into the final model which is presented in detail in Appendix C and D.
[9] A detailed description and elaboration of each project management phase can be found in the PMBOK by the PROJECT MANAGEMENT INSTITUTE (2017).

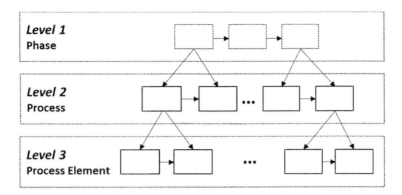

Figure 20 | Structure of the Process Reference Model

The following chapter is structured as follows: In Chapter 3.4.1, first, the phase model (level 1) is introduced. Subsequently, the individual phases are further detailed with respect to the processes (level 2) and process elements (level 3) as well as their key inputs and outputs in the remainder of Chapter 3.4.[10]

3.4.1 Phase Model

PReMMa depicts a reference process model which presents a comprehensive number of process elements. Depending on the scope of a targeted PdM solution, not all described processes and process elements need to be addressed.

In order to structure the process reference model, different phases for PdM implementation are deduced, taking into account the set of identified steps. These steps have been grouped into five distinct phases to be performed during the implementation of predictive maintenance. The resulting phases, therefore, describe the highest level of the PdM process reference model, the phase level. The phase model is presented in Figure 21, which emphasizes the relation between the phases as well as the primary outcomes.

The PdM implementation is initiated in the *preparation phase*. This phase covers all steps required to define the project scope and target. The phase is subdivided into the *project preparation* phase and the *candidate preparation* phase. The former sub-phase addresses the complete project, refines its specifications and identifies possible realizable machine candidates for PdM implementation as well as the current maintenance processes. In this context, a (machine) candidate denotes a

---

[10] An overview of all inputs and outputs per process (related to the processes on level 2) can be found in Table 31 in Appendix B as well as in the detailed process models of level 3 as depicted in Appendix D. The documents highlighted at the first and second levels of the process reference model represent the key documents that are central to the implementation of predictive maintenance.

machine that is eligible for the implementation of predictive maintenance.[11] The second sub-phase covers the planning and specification of the complete realization of a specific machine candidate. The following phases, PHM design and candidate solution deployment, are iteratively executed for each machine candidate. The system architecture design phase is executed once independently of the specific machine candidates. This phase, however, should be finalized before candidate deployment.

Under consideration of the developed concept, algorithms for fault detection, diagnosis, and prognosis are developed for each machine candidate in the *PHM design phase*. This is followed by the solution assessment with regard to predefined performance metrics and requirements, as well as the consideration of possible mitigation actions.

In parallel to the PHM design phase, the technical system architecture (technology architecture) is defined, which should enable real-time access to machine condition (sensory) data. The *system architecture design phase*, therefore, includes the physical architecture as well as the logical architecture and results in the final system implementation, including system verification and validation.

In the *candidate solution deployment phase*, the developed PHM algorithms from the PHM design phase are deployed, and a sustainment plan is elaborated. In addition, the existing maintenance processes are re-designed to fit the new requirements, and the resulting changes are subsequently realized, including corresponding personnel training and change management.

Lastly, the *project completion phase* concludes the project. This phase is carried out either as soon as all machine candidates have been processed and deployed or in case of early termination of the project. It consists of the finalization of the project documentation, the transfer of the responsibility as well as the final project assessment.

---

[11] Following the definition from Chapter 2, machine refers to the unit of analysis for predictive maintenance, which can range from individual machine components to the complete machine depending on the size, functionality and complexity of the machine.

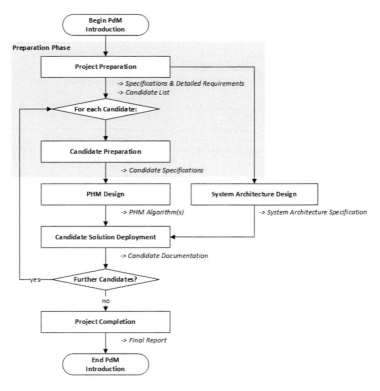

Figure 21 | Phase Model of the Process Reference Model

3.4.2 Preparation Phase

A PdM project requires detailed and comprehensive planning for its successful and target-oriented realization. For this purpose, the preparation phase depicts a critical element. The preparation phase is subdivided into the *project preparation phase*, which plans the general project scope, and the *candidate preparation phase*, which is executed for each machine candidate iteratively. The first subphase includes project specification, as-is maintenance process modeling, as well as candidate identification. In contrast, the subsequent phase covers the measurement technique selection and detailed planning. Figure 22 visualizes the process level of the preparation phase. The preparation phase consists of five processes (as depicted in the figure), each of which includes multiple process elements that are described in the following.

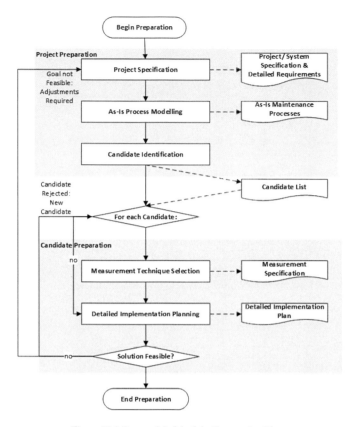

Figure 22 | Process Model of the Preparation Phase

The *project specification* in the preparation phase starts with an as-is analysis of the maintenance of the company. Based on these results, targets and objectives are defined. In addition, project stakeholders have to be identified, which range from members of the executive board to the machine operators, but can also be persons who are external to the company (e.g., customers). In particular, their expectations, their responsibilities, and potential influence on the project results have to be evaluated and addressed in the stakeholder analysis. Given both the target/ objective definition and stakeholder analysis, a general project charter (including the business case) can be derived. This is further refined in terms of project specifications and constraints, as well as requirements. The requirements analysis depicts an iterative process from high-level to low-level requirements and results in the output of the detailed requirement. As soon as the targeted project scope is defined, the economic benefit of the project has to be evaluated. In particular, the costs for PHM implementation have to be justified, and the resulting benefits identified. This is done by means of a return-on-investment (ROI) analysis. This analysis yields a final decision on the realization of the project. A critical part is the definition of appropriate key performance indicators (KPIs) (e.g., machine uptime, total maintenance costs, process quality, and error rates for RUL estimation) and envisioned targets. These should be aligned with the crucial stakeholders identified in the previous step and will provide a measure of project success. The process completes with the project plan development. The process's main output is the project and system specification, along with the detailed requirements.

The *as-is process modeling* identifies and models the existing maintenance processes, which form the basis for required process adjustments in the context of PdM. For this reason, the general maintenance processes have to be primarily understood and modeled. The modeled as-is maintenance processes will be the input for the process re-design in the candidate solution deployment phase.

The *candidate identification* targets the identification of a list of relevant PdM machine candidates, which are further considered for the PHM method design. In order to define possible PdM machine candidates, existing equipment is identified initially. From the list of equipment, potential machine candidates are subsequently assessed. This can be done in a rigorous and very structured approach, either bottom-up, via component breakdown and analysis (event tree analysis, FMECA, or hazard analysis), or top-down with regard to equipment functions using fault tree analysis, Markov/ Petri net analysis, or reliability block diagrams. All strategies result in a list of assessed machines which are evaluated with regard to their PdM necessity. The maintenance strategy determination can be performed based on the Reliability Centered Maintenance framework, as introduced in Chapter 2.1.2. The list of PdM candidates should thereafter be prioritized, taking into consideration various criteria such as availability and economic benefits. The machine candidates are iteratively taken as input to the second sub-phase, the candidate preparation, in the order of their priority.

The candidate preparation sub-phase takes as input one machine candidate. In *measurement technique selection*, initially, failure modes and required parameter measurements are identified. These are assessed against existing sensor capabilities to determine if further sensor measurements are needed. In case the available measurements are insufficient, further sensors have to be installed if the candidate allows for additional sensor mounting. Therefore, the measurement technique needs to be defined, as well as the specific measurement location identified.

Finally, *detailed implementation planning* defines and specifies the implementation of the respective machine candidate. For this reason, candidate requirements, including the targeted PHM task (fault detection, diagnostics, and/or prognostics) as well as respective performance metrics, are defined based on the detailed system requirements. The phase concludes with the final feasibility assessment and the decision about whether to continue including the candidate in the PdM project.

The preparation phase is supported by the project management tasks project initialization and planning during the project preparation sub-phase, as well as the project execution in the machine candidate preparation sub-phase.

3.4.3 PHM Design Phase

The core phase of the PdM implementation features the PHM design phase. For each machine candidate, it comprises the formulation of the required engineering knowledge, data acquisition/ preparation, algorithm development, algorithm assessment, and solution mitigation/ documentation. Figure 23 depicts the process model of the PHM design phase. The output of this phase can be either an algorithm for fault detection, diagnostics, and/ or prognostics, depending on the targeted PHM task as defined in the preparation phase. The process focuses on the development of data-driven approaches due to their high applicability in practice. Nevertheless, the process can be adjusted to physics-based approaches where more effort must be put towards understanding the degradation process rather than data preparation and algorithm development. This might be the case for safety-critical systems, where high accuracy is essential. The phase is supported by the project management tasks monitoring and controlling.

The *engineering knowledge formulation* specifies the prior knowledge about the candidate, which can be used as input for further processing by the data analysts. This process is performed without considering the data and therefore based purely on available knowledge about the machine candidate. In the first step, healthy and dysfunctional machine behavior is analyzed and described. Based on this information, targeted failure modes and associated potential parameters are selected. To evaluate the feasibility of the measurement parameters, several criteria exist, including observability, diagnosability, coverage (Goebel et al. 2017, pp. 34 f.), trendability, monotonicity, and prognosability (Coble 2010, p. 43). As soon as the parameters are defined, the candidate description is prepared, which summarizes the findings of this process to be input to the following processes.

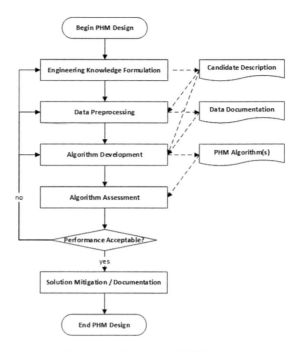

Figure 23 | Process Model of the PHM Design Phase

In *data preprocessing*, the defined data is collected, integrated, and prepared. In case a software solution is used for support, it needs to be selected and set up. The subsequent steps are carried out in alignment with an excerpt from the historical data. After data is acquired, it has to be integrated (in the case of multiple sources), reduced, and transformed. To gain an initial understanding of the data, data is subsequently analyzed through the use of simple descriptive statistics methods. In addition, data quality is verified in terms of sensor validation. This is followed by data cleaning (e.g., normalization, smoothing). Lastly, feature engineering (including feature extraction, evaluation, and selection) can be performed, and dimensionality is reduced afterward, if necessary.

The realization of the *algorithm development* depends on the targeted PHM tasks. To specify the setting, an appropriate algorithm is selected, and the test design is generated (e.g., a split of training and testing data). For condition monitoring, an anomaly or fault detection algorithm is developed, including health assessment. In addition, diagnostics covers the tasks fault isolation, failure mode identification, primary cause identification, and degradation level assessment. Prognostics, on the other hand, comprises the steps endpoint definition, health index construction, health state division, remaining useful life prediction, confidence level determination, and prognostics event horizon identification.

Thereafter, the *algorithm assessment* includes the performance evaluation under consideration of the performance metrics, as defined in the detailed implementation process, as well as the requirements verification and validation. If the objectives are not achieved, the algorithm development is re-executed.

Once the performance is appropriate, the process continues with *mitigation and documentation*. Primarily, the mitigation action is defined. Either it consists of simple maintenance instructions or complex optimization models to provide decision support to improve the planning and scheduling of the respective maintenance tasks. Based on this information, a human-machine interface or visualization dashboard is designed, and the candidate solution is documented. Lastly, a review of the performed steps in the PHM phase is conducted to identify tasks that have been overlooked as well as the determination of the next steps based on remaining resources and budget.

3.4.4 System Architecture Design Phase

The system architecture design phase describes the development of an appropriate technical infrastructure that enables continuous machine monitoring and (real-time) predictive maintenance. To that end, it is necessary to connect the machines and transfer data in a standardized manner. Control systems used for data gathering in predictive maintenance require a sound operational technology as a basis for further data processing in terms of diagnostics and prognostics. While in some cases, a sound system architecture is available, and only adaptations are required, PReMMa additionally addresses the situation with no system architecture currently in place. Beyond that, the open standards for physical asset management MIMOSA (1998-2019), which provides several industry standards with regard to enterprise application integration and condition-based maintenance, should be respected. From the project preparation phase, the phase takes as input the project specification and system constraints as well as the derived detailed (system) requirements. It comprises the processes objectives and requirements analysis, physical and logical architecture design, as well as system implementation. Figure 24 illustrates the process level of the system architecture design phase. The phase is supported by the project management tasks monitoring and controlling.

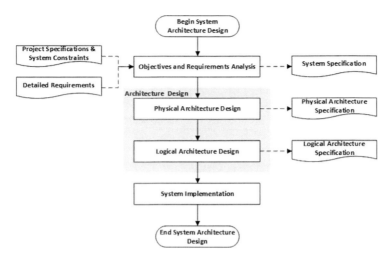

Figure 24 | Process Model of the System Architecture Design Phase

The *objectives and requirements analysis* makes use of the project specifications, system constraints, and detailed requirements that have been defined in the preparation phase. Based on these specifications, system architecture requirements are refined.

The *physical architecture design* determines the technical architecture based on the objectives and requirements. Prior to the architecture design, existing instrumentations need to be identified in order to derive further required instrumentation. Thereafter, the hardware architecture, as well as its connectivity, are specified. For this purpose, operational technology usually comprises hardware and software systems such as SCADA, DCS, or PLC, which specify connectivity and communication between the different components of the network. The detailed design of the system includes, in particular, the analysis of storage and computing options that extend or replace parts of the existing control system. In general, edge computing and cloud computing are two paradigms frequently used in conjunction with operational technology. While edge computing facilitates decentralized data processing at the edges of the network, cloud computing describes an infrastructure that is accessible via the internet either as a private (on-premise) or public (outsourced) service (Shi and Dustdar 2016, pp. 78 f.). In addition, the physical architecture design should include cybersecurity concepts. For highly sensitive data, on-premise solutions are preferred over cloud solutions.

The system architecture design is further refined with regard to the *logical architecture design*. This includes, in particular, data standardization, interface definition, and data integration. Data integration aims at integrating various data sources which are deemed as relevant for PHM design and often require integration with the company's IT systems (e.g., information available in enterprise resource planning (ERP) systems/ manufacturing execution systems (MES)). As a result of both the physical and logical architecture design, the system architecture design is defined and described.

In *system implementation*, the specified and defined system is realized and integrated within the available technical infrastructure. This is followed by the system verification (considering the system specification), transfer of responsibilities (system transition), and, as a final step, its validation.

3.4.5 Candidate Solution Deployment Phase

Once the PHM algorithm is developed and a suitable system architecture is available, the candidate solutions can be deployed. This is done either successively for each machine candidate or in batches. The candidate solution deployment phase includes deployment preparation, process adjustment, as well as candidate solution deployment, sustainment implementation, and candidate completion. Figure 25 presents the process level of the candidate solution deployment phase.

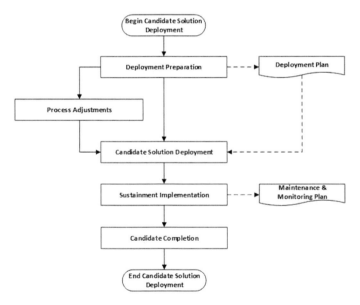

Figure 25 | Process Model of the Candidate Solution Deployment Phase

The *deployment preparation* includes the process element deployment planning with its output, the deployment plan.

Process adjustments depict the steps to integrate PdM in current maintenance business processes. Given the as-is process model developed in the preparation phase, the maintenance process of the specific machine candidate is analyzed, re-designed, and finally implemented. This is followed by personnel training. Lastly, the resulting process is assessed, considering its overall suitability, and identified adjustments are implemented. This is usually performed after the deployment of the candidate in the subsequently described parallel process.

In the process *candidate solution deployment,* the solution is deployed in terms of system integration and final candidate solution deployment. This is followed by a test and validation step. In case the testing reveals required adaptations in the system architecture design phase, the process is re-initiated.

The *sustainment implementation* targets the development of a monitoring and maintenance plan for continuous improvements of the machine candidate solution. For this reason, sustainment activities have to be defined and scheduled regularly, including the definition of effectiveness measurements, technical system review, and business process review.

Eventually, in the *candidate completion* process, documentation and review are performed, providing best practices and experiences for future machine candidate implementations. Furthermore, the responsibility is transferred to the maintenance department, and the candidate development process is reviewed.

The phase is supported by the project management tasks monitoring and controlling and the project closing for the final candidate documentation.

3.4.6 Project Completion Phase

Having implemented and deployed all PdM machine candidate solutions, the project must be formally concluded. This is done in the project completion phase. This phase should also be transgressed in case the project is terminated prematurely in order to gather the experiences and learn for future projects. The project completion phase comprises the processes completion preparation as well as the final project review. Figure 26 depicts the process flow chart for the project completion phase.

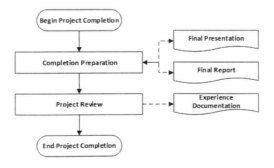

Figure 26 | Process Model of the Project Completion Phase

The process *completion preparation* targets the external completion of the project with the stakeholders. In this process, all documents are updated to the final status. Thereafter, the final project report is prepared, in general, including a written document (final report) as well as a presentation. Both documents are made available to the stakeholders and users. During a final project meeting, the results and the responsibility are transferred to the future department in charge. After the transfer, a *project review* is conducted internally with only the core project team to evaluate the overall process and project results. In that step, experiences and best practices are gathered and recorded in the experience documentation to learn for future projects.

3.5   Process Reference Model Evaluation

The final step in developing a process reference model is its systematical evaluation to assess its value and usefulness (Fettke and Loos 2003, p. 81). Process reference models can be evaluated with regards to analytical and empirical perspectives (Fettke and Loos 2003, p. 82). Analytical evaluation is performed by the designer of the model in the form of logical reasoning (Fettke and Loos 2003, p. 82). It is, therefore, time-efficient. However, it is also potentially biased towards the modeler's viewpoint (Cherdantseva et al. 2016, p. 54). The empirical perspective, on the other hand, is grounded on the experiences of potential model users (Fettke and Loos 2003, p. 82). This perspective focuses primarily on the practical value of the model and requires a significant amount of participants in order to draw informative conclusions on the model's value and usefulness (Cherdantseva et al. 2016, p. 54). In general, however, there is no superiority of one perspective over another, which advocates a multi-perspective approach to evaluation (Frank 2007, p. 123; Fettke and Loos 2003, p. 88). This multi-perspective evaluation results in a more objective assessment by benefiting from a differentiated and comprehensive evaluation (Frank 2007, p. 123). For this reason, an analytical (Chapter 3.5.1), as well as an empirical (Chapter 3.5.1) evaluation of the research results, will be conducted in the following.

3.5.1 Analytical Evaluation

For the analytical perspective of the process reference model evaluation, the descriptive feature-based approach is adopted as described by Fettke and Loos (2003). For this purpose, a feature set that specifies and characterizes reference models is used for evaluation (Fettke and Loos 2003, p. 83). In this case, the defined process reference model requirements (cf. Chapter 3.1) take over the role of the feature set for evaluation. The analytical evaluation thus aims at determining the compliance of the elaborated process reference model with the requirements that guided the development. The eight requirements are, therefore, briefly reviewed in the following. Table 5 provides an overview of the requirements and the corresponding evaluation of the process reference model.

Generality: The generality of the developed process reference model is justifiable based on the consulted literature. This can be argued from a twofold angle. Firstly, the model was synthesized from a large number of different existing process models and application cases identified through a thorough literature review. Secondly, standards were taken into account that inherently have a high degree of general validity. However, since no model is general for all cases and can cover all facets of the targeted problem, it is essential to delimit the scope for which the generality should hold (Matook and Indulska 2009, pp. 62 f.). For the developed process reference model, this applies in particular to the covered phases and considered elements. The model was developed to cover predictive maintenance projects, in particular related to the aspects of project preparation and management, data processing, system architecture design, and solution deployment. For these key aspects, the relevant process elements, the process flow as well as main data and documents were represented. Although adaptable to other settings, the process reference model was designed for production plants that target data-driven approaches within data processing.

Flexibility: The flexibility of a process reference model enables its adaptation to the individual or changing organizational requirements (Matook and Indulska 2009, p. 63). This can be realized by a reasonable structure of the process reference model as well as by its form of representation (Schwegmann 1999, p. 71). In particular, flexibility can be supported by a modular design. This entails that modifications do not have to be incorporated in the whole model but can be implemented in submodels with a lower degree of complexity (Schwegmann 1999, p. 71). For the developed reference model, modularity is achieved as a consequence of its hierarchical structure, which allows the complete predictive maintenance implementation process to be decomposed into smaller sub-processes with a greater level of detail. This enables changes to be applied to sub-processes while leaving the remaining part of the complete process unaffected. In addition, the understandability of the modeling language and conventions facilitates process modifications and therefore contributes to increased flexibility (Schwegmann 1999, p. 71). Due to the simplicity of the modeling language, as well as the syntactic ease and the limited number of symbols as defined in Chapter 3.2, this requirement is also satisfied for the developed process reference model.

Completeness: The process reference model should be complete within the given scope. This can be verified if at least one application can be demonstrated (Scheer 1999, p. 8). Analogous to generality, completeness can be justified by the literature on which PReMMa is built as well as the construction design. The huge body of literature from which the process reference model is derived thus implies high compliance with this requirement, especially considering that the integration of application processes within the case mapping and model refinement step have caused little or no changes. While completeness is fulfilled with respect to the identified knowledge base, in the second part of the assessment, conformity to practice must be verified.

Usability: The usability of the process reference model characterizes its implementability from the point of view that it is sufficiently detailed and tailored to the organizational situation (Matook and Indulska 2009, p. 63; Moody and Shanks 1994, p. 106). Consequently, it should be supportive in practical implementations. In contrast to meta-models, a process reference model should thus be applicable as a specific model (Hars 1994, p. 15). To meet this requirement, the process reference model was designed at a high level of detail containing over 100 process elements on the third level to exhaustively support the case-specific concrete design. In particular, an attempt was made to reach a balance between generality and usability, as both requirements have a negative impact on each other (Matook and Indulska 2009, p. 63). For this purpose, the phase and process level have a high degree of generalizability, whereas the process element level focuses on usability and defines potential implementation steps.

Understandability: Process reference models must be designed to be easy to understand for users (Matook and Indulska 2009, p. 62). Within this context, a highly rated attribute from the user's perspective is the simplicity and consistency of the language in use (Taylor and Bandara 2003, p. 6). This aspect was taken into account while designing the process reference model. Particular emphasis was placed on the accessibility of the language and terminology to be accessible to the model user. To this end, the terminology used in the literature was harmonized and simplified, and converted into a standardized format following the flow chart syntax defined in Chapter 3.2. In addition, a mere five modeling elements for processes and two flow elements were defined for the process reference model. Due to this limited number of modeling elements, users can easily understand the model (Schütte 1998, p. 132). Furthermore, the hierarchical structure adds clarity and transparency (Schütte 1998, p. 131). All these aspects support that this requirement is fulfilled.

Modeling Language: The modeling language should comprise all relevant elements, which are to be included in the process reference model. Furthermore, it should be matched to the needs of the users. This requirement was considered during the selection of the appropriate modeling language. For this purpose, flow charts have been chosen, which capture all relevant elements and are highly comprehensible for users without prior training.

Domain-Specific: Besides the general requirements, the process reference model for predictive maintenance should be application-independent, comprehensible, and guiding. For this reason, all phases of the project development were devised and are elaborated with a high level of detail. The model is founded on application-specific models from different areas. These models have been harmonized and generalized, making the resulting process reference model application-independent.

Table 5 | Derived Requirements for the Process Reference Model

Require-ment	Description	Resulting Characteristics
Generality	Process reference models must have a certain degree of generality. They require an abstract perspective to be applicable in different settings. To ensure general validity, the process reference model is defined for a specific scope and therefore developed for cases with comparable process characteristics.	Scope: • Data-driven • Production machinery • Process, Process flow, Data/Document
Flexibility	Process reference models must be flexibly adaptable for different situations. Adaptations and modifications for individual organizational needs are necessary. To meet the requirement of flexibility, a modular design of the process reference model is desirable.	Hierarchical process reference model
Complete-ness	Process reference models must be complete and correct. The essential process elements have to be covered, and the different situations are handled. It should, to that end, adhere to the users' requirements.	Partly covered through literature
Usability	Process reference models should represent a high degree of usability. Process reference models must exhibit a low degree of granularity to enable their implementation.	High level of detail on process element level
Under-standabil-ity	Process reference models must be simple to understand. To enable their reusability, they require high acceptance by the user. In addition, the specifics of the target group should be considered.	Modeling language and terminology
Modeling Language	The modeling language should be tailored to the scope and user requirements.	Flow charts
Domain-Specific	The process reference model should have a practical orientation. Thus, guidance should be provided throughout the implementation process, covering the majority of applications.	• Complete project scope • Great level of detail • Application-independent

3.5.2 Empirical Evaluation

While the analytical evaluation assessed the structural characteristics of the process reference model, the empirical evaluation is devoted to the content-related examination of the model and its practicality. The empirical evaluation is therefore meant to complement the subjective analytical evaluation. For this reason, the focus is primarily placed on the completeness of the model on the basis of existing predictive maintenance implementations and its reusability in the context of diverse applications. The results can be utilized to identify any subsequent need for change, which corresponds to the development and evaluation loop of design science research as described in Chapter 1.2.

In the following, first, the methodological approach for the empirical data gathering is introduced (Chapter 3.5.2.1). Thereafter, the results are analyzed and the practical conformity, as well as completeness of the model, are investigated (Chapter 3.5.2.2).

3.5.2.1 Data Gathering

The objective of the empirical evaluation is twofold. First, the fulfillment of the completeness requirement must be verified for its practical conformity. This involves investigating the suitability of the process reference model for replicating existing practical processes. From this, possible adaptation requirements for practical applications can be derived. The second objective is to measure the relevance of the individual phases and processes in practice. By this, parts of low and high practical relevance can be discovered. In contrast to the first objective, the second objective aims to identify whether the model is too complex, resulting in redundant phases and processes being represented.

Both objectives can be approached by studying existing real-world predictive maintenance implementation processes. However, in order to draw generalized and well-founded conclusions, a significant number of cases have to be examined (Dul and Hak 2008, p. 3; Cherdantseva et al. 2016, p. 54). For this purpose, multiple expert interviews are conducted, with each expert describing one case. In particular, experts from different industrial fields were consulted to capture a variety of applications, including both companies with internal predictive maintenance implementations and consultancies supporting the introduction process for their customers. The latter exhibit experience in diverse projects and have, therefore, well-established processes, which are further developed with experiences and best practices from previous projects. Table 6 depicts detailed information about the eleven industry experts, including the industry and their PdM expertise. Expert interviews are conducted in a semi-structured manner. This interview type enables room for new knowledge but also takes advantage of a predetermined structure to ensure comparability of the results to the model. The interviews are designed in consideration of the seven stages of an interview investigation defined by Kvale and Brinkmann (2009, p. 102). The interviews are structured based on four phases (Misoch 2019, p. 68). After the introduction of the participants and the interview target, the interviewees are questioned about their position and experience with PdM in the warm-up phase. The following main phase applied the dual-process mapping Widera and Hellingrath (2011, pp. 281 f.), which structures the interview in two parts (unstructured and semi-structured). Initially, in the unstructured part, the company was asked to present its PdM introduction processes without being interrupted by the interviewer. This facilitates broad information gathering. Following this, in the semi-structured part, in-depth questions about the different phases were asked to enable comparison between the various interviews. The questions focused on the steps, tasks, and documents involved.[12] To finalize, the qualitative data gathered from the interviews are analyzed using a reflexive thematic analysis. Information was extracted and structured with regard to themes enabling detailed descriptions of the presented cases (Braun et al. 2019, pp. 852–857). These were derived directly

---

[12] The guideline for the expert interviews is provided in Table 32 in Appendix E.

from the questions and the theoretical background. In addition, a few codes were added based on the qualitative data, in particular for specific process steps highlighted during the unstructured process narration. Finally, the processes were reviewed again by the experts to ensure that the interview content was captured correctly and to eliminate any possible misunderstandings.[13]

Table 6 | Expert Information

Company	Industry	Experience
A	Steel Tube Manufacturer	1-2 years
B	Automotive Manufacturer	1 year
C	Aircraft Engine Manufacturer	2 years
D	Aircraft Manufacturer	3 years
E	Wind Turbine Manufacturer	4 years
F	Research Institute	3-4 years
G	Management Consultancy	3-4 years
H	Management Consultancy	2-3 years
I	Management Consultancy	10+ years
J	IT Service Provider + Consultancy	4-5 years
K	Solution Provider	11 years

3.5.2.2   Data Analysis

Given the identified processes, in a first step, the fulfillment of the completeness requirement is evaluated. For this purpose, the identified practical processes are individually matched against the derived process reference model. Each identified step was examined to assess whether it is reflected in the process reference model. Moreover, the process and information flows were benchmarked against the flows specified by the process reference model.[14]

The final results of this evaluation step demonstrated that, with minor exceptions, the practical processes could be mapped to the process reference model. Thus, although only a subset of the theoretically-derived steps is implemented within each of the eleven cases, it is possible to show its conformity with the developed process reference model as well as the completeness of the developed process reference model. Nevertheless, the analysis has shown that minor modifications of the process reference model can improve the quality of the model to better meet the requirements arising from industrial applications. This comprises the extension of the process reference model by fifteen process elements within different phases and processes, most of which were mentioned by several experts. In addition, the candidate preparation phase could be better structured, in particular, taking into account the additionally identified process elements. For this reason, there should be a preparatory step in the candidate preparation phase to identify and determine possible preconditions to allow for early termination if the solution is

---

[13] The short description of the identified practical processes as well as the summary of the identified steps by each interview are provided in Appendix F and G.
[14] The results of their individual assessment in terms of their fit to the process reference model is provided in Appendix F after the corresponding process description.

not feasible. Table 7 provides an overview of the resulting need for adaptation of the process reference model in terms of missing steps as well as the reference to the practical process that indicated the need.

Table 7 | Industry Process Assessment

Phase	Processes	Need for Adjustments based on Expert Interviews	Co.
Preparation	Project Preparation	Maturity Assessment	I, J
	Candidate Identification	Component Definition (Criteria-based Approach)	C
	Prerequisite Determination	Team Formulation, Pre-Feasibility Analysis	D, E, F, J
	Measurement Extension	Measurement Market Research, Measurement Installation, Measurement Testing	A
	Concept Elaboration	End-to-end Use Case Definition, Concept Development	I, B
System Architecture Design	Objectives and Requirements Analysis	Operationalization Validation	G, H, J
	Physical Architecture Design	Market Review	B
PHM Design	Data Pre-Processing	Data Visualization/ Data Analysis, Data Labeling	A, B, C, F, G
Candidate Solution Deployment	Deployment Planning	Industrial Implementation	G, D
	Process Adjustments	Personnel Training/ Change Management	C

The second objective is addressed by investigating the coverage of the phases and processes by the different industrial cases. For this purpose, it is examined which phases and processes of the developed process reference model have been carried out by the companies in order to draw conclusions on the general relevance of the phases and processes. For the purpose of quantifying the relevance of the individual phases and processes, two measures are specified, namely the Phase Coverage Percentage (PhCP) and the Process Coverage Percentage (PrCP). The Phase Coverage Percentage depicts the percentage of companies that have addressed the respective phase. This measure is therefore defined as:

$$PhCP = \frac{number\ of\ cases\ covering\ the\ phase}{total\ number\ of\ cases} \tag{3.1}$$

Analogous to the PhCP measure, the Process Coverage Percentage represents the percentage of companies that mentioned the analyzed process within the respective phase. The Process Coverage Percentage is calculated as follows:

$$PrCP = \frac{number\ of\ cases\ covering\ the\ process}{number\ of\ cases\ covering\ the\ phase\ of\ the\ process} \tag{3.2}$$

While the PhCP gives an indication of the relative importance of the phase, the PrCP characterizes the relevance of the process within the phase.

Results of the practical phase and process relevance are depicted in Table 8. Regarding the phases represented by the PhCP measure, it can be stated that the phases of preparation, PHM design, and candidate solution deployment are executed within all cases and are therefore essential within a predictive maintenance project. The phase system architecture design is targeted by almost half of the cases and is only required if the pre-existing architecture is not sufficient for predictive maintenance. In addition, the project completion phase was mentioned only by two companies. However, this aspect could be explained by the fact that, in most cases, the implementation has not yet been fully completed. The PrCP measure, on the other hand, serves as an indicator of the importance of the respective process within the phase. Processes with a high coverage percentage should at least be considered in detail and be evaluated for the specific case. It can be shown that the vast majority of processes have been targeted by half of the practical cases. Only two processes are not addressed by any of the eleven cases. These processes are the as-is process modeling as well as the project review. The former can be attributed to the fact that the interrogated companies already have a good understanding of their current maintenance processes, while the latter is due to the overall low coverage of this phase, as most of the projects have not yet reached completion. However, for these reasons, the lack of coverage does not lead to the exclusion of these processes within the process reference model.

Table 8 | Process and Phase Coverage

Phase	Processes	PhCP	PrCP
Preparation	Project Preparation	1.0	0.73
	As-Is Process Modeling		0
	Candidate Identification		1.0
	Prerequisite Determination		0.45
	Measurement Extension		0.45
	Concept Elaboration		0.18
System Architecture Design	Objectives and Requirements Analysis	0.45	0.60
	Physical Architecture Design		0.20
	Logical Architecture Design		0.20
	System Implementation		1.0
PHM Design	Engineering Knowledge Formulation	1.0	0.18
	Data Preprocessing		0.91
	Algorithm Development		1.0
	Algorithm Assessment		0.73
	Mitigation & Documentation		0.64
Candidate Solution Deployment	Deployment Preparation	1.0	0.09
	Process Adjustments		0.55
	Candidate Solution Deployment		0.91
	Sustainment Implementation		0.63
	Candidate Completion		0.18
Project Completion	Completion Preparation	0.18	1.0
	Project Review		0

To sum up, within the first part of the empirical evaluation, it was shown that the process reference model is well-aligned with the practical processes. Only a minor need for adjustment was identified. A similar result was reached within the second part of the empirical evaluation. Within this evaluation, it was demonstrated that the practical processes apply all of the phases and the vast majority of processes. Both evaluations support the general claim for the reusability of the process reference model and its practical conformity, given the diverse set of applications considered.

The resulting minor adjustments from the first part of the empirical evaluation have been included in the process reference model in order to improve its quality and make it more applicable. The resulting process reference model is provided in Appendix Section C and D.

3.6 Discussion and Limitations

PReMMa is a process reference model that supports the realization of an industrial predictive maintenance project. It provides guidance and support for the design of a predictive maintenance implementation process within all phases and tasks to be performed on three different levels of detail. PReMMa is designed to be universally applicable and is adaptable to the characteristics and scope of a specific project. The process reference model includes the reference process as well as key input and output documents, which are created and used. It is grounded on theoretical knowledge, which is validated by industry insights with regard to eleven expert interviews. The resulting needs for adjustment have been incorporated into the model, following the development and evaluation loop of the design science research methodology.

The process reference model is founded in and backed by a large body of literature. This is a great strength in terms of rigorous development but at the same time a limitation. This can be ascribed to the fact that the material under investigation is not fully comprehensive and is limited by the identified models from the literature. In particular, due to the existing inconsistent terminology within the target field and the resulting need to adapt the literature search procedure, complete coverage of literature was not attainable. While a large number of publications and standards are identified, the literature search process remained at the level of representative coverage as classified by Cooper (1988, p. 111). Nevertheless, it can be argued that the identified knowledge base was sufficiently large to thoroughly capture the predictive maintenance implementation process. This is supported both by the size and level of detail of the model and by the fact that models from the literature included in the final stages of the synthesis resulted in little to no change to the process reference model. Moreover, the empirical evaluation has revealed a high degree of compatibility with the developed process reference model. Resulting minor adjustments were implemented to improve the general validity of the process reference model and to overcome the challenge of rigorous development. Future findings and results should be included to enable continuous improvement of the process reference model. In particular, existing industry challenges could be taken up and better accommodated within the model. In addition, further evaluations of the process reference model could improve its overall quality.

Furthermore, due to its general validity, the developed process reference model is not exhaustive regarding specific application-dependent requirements and challenges. It provides a structured approach that outlines the essential processes. When designing the model, a trade-off between generalizability and detailed application-specific implementation had to be reached. This has resulted in the first two levels of the process reference model being very general and universal. In most cases, however, the detailed steps on the third level rather serve for orientation purposes and have to be adapted considerably in order to be used in a specific application. Considerable effort is therefore required to adapt the generic model to the specific application requirements. Nevertheless, the process reference model can greatly facilitate this task. For improved coverage, the further development of the process reference model would have needed to address more specific application areas, albeit limiting the generalizability. Finally, the process reference model is limited to its defined scope. This applies in particular to the considered elements. The emphasis was placed on the processes, the process flow, and the corresponding documents. Further perspectives for improved support should be investigated and elaborated. This could include, among other aspects, the involved persons and project team as well as the concrete design of the individual processes.

While the process reference model is designed to be application-independent, the fleet prognostics development method is focused on the specific use case of processing fleet data. For the subsequent part of this work, it is, therefore, necessary to determine whether the process reference model needs to be adapted for the purpose of fleet prognostics and, if so, which processes need to be modified. As the fleet prognostics method aims to develop a prognostic algorithm based on fleet data, the PHM design phase is of utmost importance, especially in terms of data preprocessing, algorithm development, and algorithm evaluation processes. Within these processes, the handling of heterogeneous data requires a more detailed examination. For the remaining phases of the process reference model, however, there is no need for adjustment in order to implement the fleet prognostics. This reasoning can be substantiated by both the literature and the results of the expert interviews. From the conducted literature search, only the paper by Johnson (2012) was identified that addresses the processing of fleet data. The author defines a four-step design process for fleet-wide asset monitoring. These steps are found in the majority of identified processes in the literature and therefore do not represent a specification of the process reference model for fleet prognostics. However, it is emphasized that an appropriate selection of machines within the fleet must be performed. In addition, the experts who were questioned during the empirical evaluation were asked whether the inclusion of fleet data had resulted in any changes to their process. In response to this question, the experts unanimously responded with the algorithm development process, excluding other processes. Given this information, no changes or adjustments are required for the process reference model in the context of fleet prognostics, with the exception of the detailed elaboration (level 3) of the PHM design phase. However, the machines that compose a fleet should be defined as well as the data that is available for analysis. Hence, only the detailed design of the identified data preprocessing, algorithm development, and algorithm evaluation processes within the PHM design phase will be researched for the implementation of fleet prognostics in the following two chapters. To this end, a fleet prognostic development method is provided, which supplements the process reference model with respect to these three processes.

4 Characterization Method for Fleet Prognostics

The above-developed process reference model defines five phases to realize predictive mainte-nance in a company. The key phase in PdM projects is the PHM phase, which is dedicated to the development of algorithms for fault detection, diagnostics, and prognostics. When algo-rithms are implemented in a data-driven manner, this phase highly depends on the availability of historical data. While data is often limited to individual machines, knowledge from a fleet of machines can be used to increase its volume. However, due to the heterogeneous nature of machines within a fleet, this comes with various challenges. To adequately address the hetero-geneity of machines in the PHM phase, the fleet needs to be characterized adequately to take the multiple sources of heterogeneity between machines into account for the development of algorithms. For this purpose, the process reference model is supplemented by a method for fleet characterization.[15] The method focuses on fleet characterization for prognostics.

This chapter thus presents the developed method for fleet characterization for prognostics. For this purpose, primarily, objectives and requirements (Chapter 4.1) are elaborated, and the meth-odological approach (Chapter 4.2) is presented. Thereafter, the basis for the fleet characteriza-tion method is laid by providing the consolidated knowledge from literature (Chapter 4.3). This is followed by the presentation of the elaborated characterization method for fleet prognostics (Chapter 4.4). To conclude, the developed method is evaluated using experiments (Chapter 4.5), and finally, results are discussed (Chapter 4.6).

4.1 Objectives and Requirements Definition for the Characterization Method

Data-driven approaches require a large amount of data to fully understand and learn the degra-dation behavior of the system under investigation. While this volume is often not available for single machines, data obtained from a fleet of similar machines can be considered to increase its representativeness (Michau and Fink 2019, p. 2). This enables to capture further potential behavior patterns and learn from other machines in an intelligent way (Fan et al. 2015, p. 448). The increased data volume can lead inter alia to improved long-term accuracy (Liu et al. 2007, p. 558) and robustness (Lapira 2012, p. 56). However, data from a fleet of machines is hetero-geneous in nature due to unique degradation patterns as a result of usage and environmental exposure (Bonissone et al. 2005, p. 20). If used with traditional data-driven approaches, though, only global degradation behavior is gathered (Liu and Zio 2016, p. 1). The systematic consider-ation of differences between machines within prognostics has only received increased attention in recent years and has therefore not yet been treated exhaustively in the literature (Zhou et al.

---

[15] Preliminary results of the fleet description as well as parts of the state-of-research of this chapter have been published in the Proceedings of the Annual Conference of the Prognostics and Health Management Society in the paper "Fleet Knowledge for Prognostics and Health Management — Identifying Fleet Dimensions and Characteristics for the Categorization of Fleets" (2017) by Wagner and Hellingrath.

2016, p. 13). Current research targets application-specific approaches by addressing their unique characteristics. Depending on the specific application, however, a fleet may exhibit a variety of characteristics that have different implications for the inclusion of its data for prognostics. Therefore, to enable a more generic and application-independent view on fleet prognostics, it is first required to determine the range of fleet characteristics as well as their implications for prognostics. From this, it is possible to classify fleets accordingly and derive their imposing requirements. In this context, the characterization of a fleet depicts the identification and description of its distinctive features with regard to fleet prognostics. The characterization can be accomplished with the help of a method. The method to be developed should therefore enable a fleet to be characterized on the basis of its distinctive features with the aim of deriving fleet-specific requirements for prognostics. Thus, for the development characterization method, three requirements can be stated:

Fleet Characteristics: The characterization method should be designed to facilitate a structured identification and description of relevant and differential characteristics of fleets for the purpose of prognostics. In this way, the method should provide a better understanding of the fleet.

Fleet Prognostic Requirements: The characterization method should enable the determination of requirements for fleet prognostics. These requirements should facilitate the development of an eligible prognostic algorithm, taking into account the fleet-specific characteristics.

Characterization Process for Fleet Prognostics: The characterization method for fleet prognostics must provide a systematic process that captures a set of activities and their sequence. This process should be adaptable to the specific characteristics of the fleet and its data by providing recommendations for the concrete design of its activities (Winter et al. 2009, p. 3).

The first two functional requirements can be derived from the application domain, while the third structural requirement arises from the targeted artifact (method).

The first requirement entails a conceptual view of fleet prognostics. As the fleet can have various manifestations, it is necessary to study the different potential characteristics and distinct properties within fleet prognostics. On the one hand, this can clarify the distinction from prognostics in general and identify the unique fleet prognostic characteristics. On the other hand, it contributes to a comprehensive understanding of the fleet.

Given the set of identified fleet characteristics, the second requirement translates the identified fleet characteristics into fleet prognostic requirements. A data-driven perspective is adapted for this purpose. In particular, this demands the definition of fleet prognostic types from which specific fleet prognostic requirements can subsequently be derived. Here, a fleet prognostic type represents a group of fleets with similar and comparable data characteristics and thus degradation and failure behaviors. Furthermore, the adapted data-driven perspective allows quantifying the impact of the fleet characteristics on the data, enabling an improved characterization of the fleet.

The identification of the corresponding data-driven fleet prognostic types as well as the fleet characteristics should be supported by a method, which is addressed by the third requirement. The method comprises both a process (including activities and their sequence) as well as recommendations with regard to the design of the activities. By this, the user should be guided to properly characterize their fleets and derive fleet prognostic requirements. The resulting method should, thus, enable a structured and target-oriented analysis of the fleet and its data as well as derive fleet prognostic requirements, which can subsequently be used within the algorithm development method for fleet prognostics. Given the requirements derived through the characterization method, possible appropriate fleet prognostic methods are proposed, and their development assisted as part of the algorithm development method in Chapter 5.

4.2 Methodological Approach

The fleet prognostics characterization method is developed with regard to the *seven-step method development process* depicted in Figure 27. It aims at gaining a deeper understanding of the fleet concept and its usage within prognostics, taking into account a fleet's diverse characteristics, and from this, derive a method for fleet prognostics characterization. The methodological approach, therefore, addresses the specified requirements in a structured manner. Literature research is used as a foundation for the development of the method. It covers the second type of literature review, as defined by Webster and Watson (2002). In this case, the literature analysis addresses a newly evolving topic and thus, aims at deriving a theoretical framework using the information obtained in the literature. In comparison to literature research in well-established topics, these reviews are usually shorter and establish a knowledge basis for further development (Webster and Watson 2002, p. xiv).

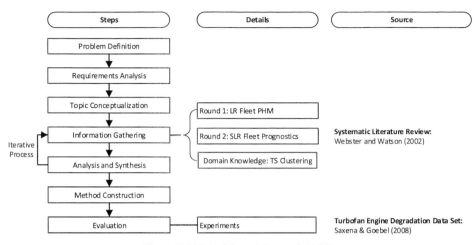

Figure 27 | Methodological Approach (RQ2)

The performed methodological procedure is motivated by the *framework for literature reviewing* (vom Brocke et al. 2009) as well as the *process of writing a review article* (Webster and Watson 2002) and is furthermore extended for the method development with the step method construction in accordance with the design science methodology (cf. Chapter 1.3). In order to define the scope of the literature research and the method, the approach starts with a clear definition and delimitation of the problem and, building on this, deduces requirements for the method (cf. Chapter 4.1). This is followed by a presentation of the current state of the art and existing shortcomings in the topic conceptualization (vom Brocke et al. 2009, p. 8). This step encompasses a working definition of key concepts (Zorn and Campbell 2006, p. 176; vom Brocke et al. 2009, p. 8). In this case, the fleet definition for prognostics is investigated in detail. Given the general introduction to the topic, information is gathered through literature research and subsequently analyzed and synthesized. The results are subsequently used for the construction of the fleet characterization method. In the course of the information analysis and synthesis, it was identified that time-series clustering depicts a suitable option to assess fleet data from a data-driven perspective and derive fleet prognostic requirements. For this purpose, concepts of time-series analysis are transferred to the domain of fleet characterization. The resulting method is evaluated considering experiments on a well-established turbofan engine degradation data set.

The identification of suitable literature for fleet diagnostics and prognostics is demanding due to non-standardized terminologies as well as the multi-disciplinary research field. For this reason, an iterative search process with two rounds was chosen for the information gathering step. The first round targeted a non-systematic literature review in the general field of PHM with fleet data with the objective of identifying some representative literature and providing a broad view of the research field. In particular, initial findings are attempted to establish a good foundation and refine the scope for the second, more in-depth analysis. Thus, due to its simplicity and time efficiency, the approach of the (narrative) literature review was targeted (Paré et al. 2015, p. 185).

It does not aim at the generalization of the results; however, it depicts a rather opportunistic survey of easily accessible literature in order to summarize and build upon prior findings (Paré et al. 2015, p. 185; Grant and Booth 2009, p. 94). The literature search is, therefore, neither exhaustive, comprehensive, nor systematic (Paré et al. 2015, p. 185). It intends to identify key characteristics (Grant and Booth 2009, p. 94), which are exploited in the second round of the literature review process. In addition, this approach enabled a search process with a certain degree of impartiality, which is not narrowed down to a concrete search design. For this, related terms of the fleet (e.g., fleet, similar machines, groups of machines) and predictive maintenance (e.g., prognostics, diagnostics, condition monitoring) are searched in various databases, and thirty promising articles are selected for deeper investigation. These publications are used to develop a general fleet description by identifying dimensions and respective characteristics of fleets to be considered in the context of PHM with fleet data.

The developed fleet-wide PHM ontology by Monnin et al. (2011b) and Medina-Oliva et al. (2014) is used as a foundation for the structuring of the identified characteristics. The ontology enables the classification of machines within a fleet with regard to five different contexts, namely technical, service, operational, dysfunctional, and application context (Medina-Oliva et al. 2014, pp. 44 f.; Monnin et al. 2011b, p. 4). These contexts are used as a starting point. In an iterative process, they are complemented and detailed with the categories and corresponding character-istics provided in the literature. Due to the high importance of the available data for diagnostic and prognostic approaches, the data dimension is additionally included. This process resulted in a total of four fleet dimensions, namely the application area (extension of the application context), the fleet composition (synthesis and detailing of the technical and dysfunctional con-text), the operating condition (synthesis and detailing of the service and operational context) as well as data.[16]

The initial, rather broad and non-systematic, literature review provided a preliminary under-standing of the fleet and its key characteristics for PHM applications. Based on these results, the heterogeneity of machines within the fleet was identified as a primary differentiator in the usage of fleet data for PHM (cf. Chapter 4.3.2). For this purpose, the second round of literature analysis targeted a more in-depth analysis of the heterogeneity of machines. In particular, the various origins of heterogeneity among machines are identified, and their effects on degradation behavior assessed. This analysis enables an improved characterization of the fleet and, thus, a more targeted development of a fleet prognostic algorithm. In order to identify the causes of fleet heterogeneity, a second literature review was performed, which aims to extend the results of the first literature review. Analogous to Chapter 3.2, the methodology by Webster and

---

[16] The dimensions, their categories as well as the corresponding characteristics will be presented in Chapter 4.3.2 as well as in Chapter 4.4.1 as part of the general fleet description.

Watson (2002) was considered as it provides a structured approach for a literature review without being too restrictive. In particular, it aims at identifying a strong foundation for the elaboration of the topic without seeking completeness (Webster and Watson 2002, p. xvi). Due to its rather flexible procedure, the approach can be adapted to the problem at hand. For this purpose, the search process for literature was twofold. On the one hand, a plurality of synonyms for fleets as well as for prognostics are searched, and a forward and backward search is applied subsequently. As the first keyword search revealed only 18 publications, the search process was adapted to include further publications which considered the established C-MAPPS data sets because these model the degradation of turbofan fleets. Multiple C-MAPPS publications are identified through the review paper by Ramasso and Saxena (2014), as well as by an additional forward and backward search. The two searches resulted in a total of 26 papers, which are analyzed to identify specific sources of heterogeneity within fleets.[17]

4.3 Fleet Conceptualization for Prognostics

In order to better characterize the fleet and its data, the term fleet needs to be analyzed from a conceptual perspective and key characteristics elaborated from the literature. Therefore, this chapter provides the foundation for the development of the characterization method for fleet prognostics. In Chapter 4.3.1, the fleet is defined, and existing concepts are presented. Subsequently, in Chapter 4.3.2, the results from the literature review are summarized in terms of general fleet dimensions, with a focus on the heterogeneity of machines within a fleet. Building on these results adapted and refined fleet prognostic types are introduced in Chapter 0, which depict the analysis and synthesis step of the methodological approach. Lastly, Chapter 4.3.4 provides a discussion on the identification of the derived fleet types using time-series clustering.

4.3.1 Fleet Definition

The word fleet is a term widely used and known in different areas. Nevertheless, no common and concise understanding is available in the literature. This is mainly attributed to the different contexts it is applied. In most cases, the term fleet is associated with a specific type of fleet having unique characteristics. While fleet in fleet management mainly refers to managing the transportation fleet (e.g., vehicle, ship) of a company, fleets also exist in other contexts like industrial systems (Monnin et al. 2011a, p. 3). Depending on the specific context, the fleet can have different characteristics and features. Common dictionaries define the term fleet as a group *"operated under unified control"* (Merriam-Webster Dictionary) or *"engaged in the same activity"* (Oxford Dictionaries). These definitions are rather vague and mainly focus on fleets of ships. With regard to PHM, the term fleet is often used without a detailed description, mainly presenting the fleet

---

[17] The exact keywords used for the initial search and the final composition of the identified body of literature are outlined in Figure 73 in Appendix A.

dimension as an additional data source (e.g., Saxena et al. 2008b, Fan et al. 2015) and generally referring to the fleet as a set of similar machines (e.g., Al-Dahidi et al. 2017, p. 2). Only a few publications provide a detailed specification of their understanding and the composition of the machines within the fleet.

In general, two different predominant definitions can be deduced from the available literature within PHM. On the one hand, a fleet refers to a set of machines (either systems, sub-systems, or equipment), which are grouped together for a specific purpose, a given time and share some characteristics (Monnin et al. 2011a, p. 3). In this definition, fleets depict the set of machines (or a subset) belonging to one owner (Medina-Oliva et al. 2012, p. 2; Jin et al. 2015, p. 225). On the other hand, a fleet is regarded as a set of homogeneous machines with compatible characteristics, however, exposed to different operating conditions (Michau et al. 2018, p. 2). This definition does not restrict the fleet to one single operator. While the first definition is closely related to the operators' perspective on fleets, the second definition focuses more on the manufacturers' perspective (Michau et al. 2018, p. 2). Both fleet definitions are, however, subject to a broad understanding of the fleet and may, therefore, result in a very inhomogeneous set of machines (Michau et al. 2018, p. 2). For this purpose, it may be required to identify similar or identical sub-fleet based on technical machine characteristics (Monnin et al. 2011a, p. 4), similar degradation patterns (Leone et al. 2017, p. 164), with respect to common semantics (Medina-Oliva et al. 2014, p. 43) or categories of interest (Cristaldi et al. 2016, p. 2).

Taking these definitions into account and aligning them, a fleet can be described as a *"set of machines, which are linked with regard to similar characteristics"*. In the context of PHM, the fleet definition must be further complemented by the aspect that machines in the fleet or sub-fleets are *assumed to have similar failure indicators, types, or degradation behaviors*. Nevertheless, machines within a fleet can be exposed to various different operating conditions. While this working definition of a fleet in the context of PHM provides a good basis, it does not concretize the degree of similarity. For this reason, further investigations on machine similarity will be conducted in the following chapters.

4.3.2 Information Gathering

To gain a good understanding of the current fleet in the context of PHM, it is useful to specify its general characteristics in a structured way. With respect to the literature identified within the first literature review round, a fleet in the context of PHM is usually described with regard to four dimensions, namely its application area, the fleet composition, its operating conditions, and the available data:

Application Area | The application area describes and specifies the general information on the fleet, respectively, the set of machines. In particular, the unit of analysis is defined as well as the targeted task. The same application area for all machines in the fleet is a prerequisite for its further usage within PHM. For example, it will not be possible to treat trains and aircraft within one fleet in PHM; however, different types of aircraft could be eligible for a fleet approach as they reflect the same area.

Fleet Composition | The fleet composition specifies the structure of the fleet, in particular in terms of the diversity of the machines within the fleet. The describing criteria include, among others, the size of the fleet, differences in the system design, as well as age structure. In particular, the heterogeneity of the machines in terms of their structure and functionality is explained. This dimension helps to better detail the composition and thus conceptually defines the degree of machine similarity.

Operating Condition | The operating condition reflects the operating and environmental parameters to which the machines in the fleet are subjected. In particular, it captures the extent of disparity in the usage, load, and environmental condition of the machines in the fleet. Given that machine degradation correlates with their operating conditions, this dimension is decisive for fleet prognostics to account for possible discrepancies between machines within their operating parameters when developing a prognostic algorithm.

Data | The data dimension describes the data and its structure. This dimension is helpful both to gain a better understanding of the data and to explicitly consider data characteristics as part of the prognostic algorithm. The data dimension can be grouped further into data description, data properties, and data transmission. The data dimension is, to a large extent, valid for all PHM applications. Although it could be possible that the defined data characteristics differ between machines (e.g., transmission frequency, dimensionality), these dissimilarities have to be accounted for during data preparation. For a fleet PHM application, it must therefore be ensured in advance that the data characteristics are comparable. In case this is not possible, fleet data usage is heavily restricted and, in most cases, not possible.

Table 9 depicts the concept matrix of the identified publications and their contribution to each dimension. In addition, for the application area dimension, the detailed specifications of the literature are provided.

Table 9 | Concept Matrix – PHM with Fleet Data

Sources	Application Area		Fleet Composition	Operating Condition	Data
	Unit of Analysis	PHM Context			
Agarwal et al. 2015; Agarwal et al. 2012	Nuclear Power Plants	Diagnostics and Prognostics (Architecture)		X*	X*
Al-Dahidi et al. 2016	Automotive Vehicles/ Aircraft Engine	Prognostics		X	X*
Bagheri et al. 2015	Aircraft Engine	Prognostics		X*	X*
Bonissone and Varma 2005	Locomotive	Mission Reliability		X	
Byttner et al. 2011; Fan et al. 2015	Bus	Diagnostics	X*	X*	X*
Fang et al. 2010	Aircraft Engine	Reliability Assessment		X	X*
Gebraeel 2010	Aircraft Engine	Mission Reliability	X*		
Guepie and Lecoeuche 2015	Railways	Prognostics			X*
Jin et al. 2015	Railways	Framework		X	
Krause et al. 2010	Manufacturing	Architecture	X*		
Lapira 2012	Wind Turbines/ Industrial Robots	Fault Detection	X*		
Le and Geramifard 2014	Milling	Health Assessment		X	X
Léger and Iung 2012	Ships	Monitoring; Detection; Prognostics	X*	X	X
Leone et al. 2017	Ships	Prognostics			X*
Liu and Zio 2016	Nuclear Power Plants	Prognostics	X	X*	X
Liu et al. 2007	Production Machinery (spindle)	Diagnostics; Prognostics		X	X*
Medina-Oliva et al. 2014; Medina-Oliva et al. 2012; Monnin et al. 2011a	Ships	Prognostics	X*	X*	X*
Patrick et al. 2010	Helicopter	Diagnostics	X	X	
Saxena et al. 2005	Automotive Vehicles	Diagnostics	X	X	X*
Schneider and Cassady 2004	- Not specified -	Selective Maintenance	X*	X*	X
Shumaker et al. 2013	Nuclear Power Plants	Prognostics		X	
Subbiah and Turrin 2015	Circuit Breaker	Prognostics		X	(X)
Xue et al. 2008	Aircraft Engine	Prognostics	X	X	X
Zaidan et al. 2016	Aircraft Engine	Prognostics	X	X	X
Zio and Di Maio 2010	Nuclear Power Plants	Prognostics	X		X
Zuccolotto et al. 2015	Electric Valve	Diagnostics	X		X

Legend: (*) indicates the main contributors with respect to characteristics and distinct attributes within the dimension

From the identified dimensions and their discussions, it can be deduced that the dissimilarity of machines within a fleet depicts the key characteristic, which makes fleet prognostics more challenging compared to general prognostic applications. The identified categories with the dimensions fleet composition and operating condition provide a good foundation; however, they are not exhaustive. For this purpose, the fleet prognostic literature is analyzed, and further sources of heterogeneity between machines are investigated (cf. Chapter 4.2). Figure 28 highlights the identified sources of machine dissimilarities for fleet prognostics.

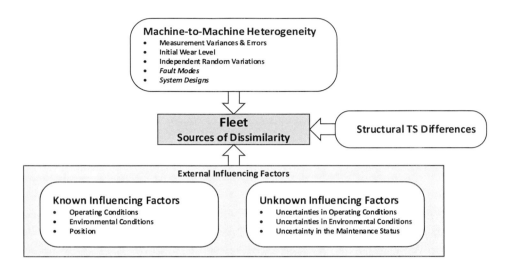

Figure 28 | Fleet Heterogeneity – Detailed

From the body of literature (second round of the literature review), three main sources of machine dissimilarity are determined, namely machine-to-machine heterogeneity (including different fault modes), external influencing factors (both known and unknown), and structural time-series differences.

Machine-to-Machine Heterogeneity | In general, machines within a fleet exhibit machine-to-machine heterogeneity. Even though operated in the same manner and in the same environment, machines will evolve differently. This is, on the one hand, attributed to independent random variations as well as measurement variations. Apart from that, machines might exhibit varying initial wear levels or various system designs. The latter can have various forms; among them are system variants or customized systems, different numbers of sensors, or variations caused by manufacturing or material. An additional source of heterogeneity is multiple fault modes. These can either be labeled (known) or unidentified.

External Influencing Factors | Besides machine-to-machine heterogeneities and differing fault modes, the degradation of machines is influenced by several external factors, which are separated into known and unknown factors. Even though there might be more external influencing factors, the most frequently considered are operating and environmental conditions. The information can only be available explicitly (known) or in the form of uncertainties (unknown – non-existent). Further, different physical positions of components within a machine (e.g., brake pads of a train) were used as a known influencing factor, whereas the actual maintenance status of the machines is specified as an unknown influencing factor.

Structural Time-series Differences | Lastly, structural differences in the form of irregular sampling intervals are discussed. While this aspect does not directly affect the degradation behavior, however, it has to be considered for fleet prognostics during data preparation and is therefore included.

Table 10 | Concept Matrix – Sources of Dissimilarity in Fleet Prognostics

	Machine-To-Machine Heterogeneity					Known Influencing Factors			Unknown Influencing Factors				Others
	Measurement Variances	Initial Wear Level	Independent Random Variants	Fault Modes	System Design	Operating Conditons	Environmental Conditions	Position	Uncertainty in the Environmental Condition	Uncertainty in the Operational Condition	Unidentified Failure Modes	Uncertainty in the Maintenance State	Irregular Sampling Time
Al-Dahidi et al. 2016					X	X							
Babu et al. 2016		X				X							
Bektas et al. 2019		X				X							
Chang et al. 2017				X					X	X	X		
Cristaldi et al. 2016													
Duong et al. 2018	X			X		X			X		X		
Frisk and Krysander 2015						X	X						
Gebraeel et al. 2004	X		X						X				
Hu et al. 2012									X	X			
Lam et al. 2014						X							
Liu and Zio 2016						X	X						
Palau et al. 2020			X										
Peng et al. 2012			X	X		X							
Ramasso 2014			X	X		X					X		
Riad et al. 2010		X											
Rigamonti et al. 2015						X			X				
Song and Lee 2013						X	X						
Trilla et al. 2018								X					
Voisin et al. 2013					X	X	X						
Wang et al. 2008						X							
Wang et al. 2018					X				X				
Wu et al. 2017b						X							
Yang et al. 2017						X	X						X
Yu et al. 2012						X							
Zaidan 2014; Zaidan et al. 2015a		X			X	X			X			X	

clearly separable degradation behavior is observed in terms of distinct sub-fleets. This fleet behavior may arise from limited discrete values for external factors with a high impact on the degradation trajectory.

System specifications and fault modes constitute an exception in this fleet type definition with regard to their classification as a machine-to-machine heterogeneity source of dissimilarity (cf. Chapter 4.3.2). Different system specifications could have a greater impact on degradation behavior, which cannot be described as noise. The effect, however, highly depends on the specific form of variations. The larger the differences in the system design, the greater the effect on the degradation behavior. For this purpose, system specifications should be treated as external influencing factors within the assignment of fleet prognostic types. The same applies to fault modes. Here, different fault modes will result in different degradation behavior and are thus handles as external influencing factors.

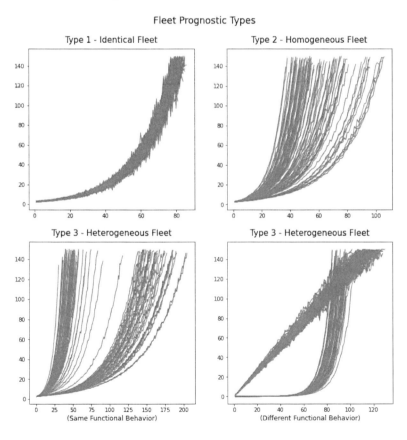

Figure 29 | Fleet Prognostic Types (Examples)

Figure 29 shows an exemplary visualization of the different fleet prognostic types based on simulated data taking into account the respective described characteristics. In particular, two different simulations were performed for the heterogeneous fleet, one with the same functional behavior of the degradation curve and one with different functional behavior. It can be seen that machines in the identical fleet exhibit the same functional behavior and only differ with regard to noise. The homogeneous and heterogeneous fleets are characterized by additional external influence modeled as random variations in the model parameters or different functional degradation behavior. While for the homogeneous fleet, no distinct clusters are visible, it is evident for both heterogeneous fleets.

The fleet prognostic types enhance the conceptually determined differences of machines in a fleet from a data-driven perspective, from which requirements for prognostic algorithm development can be derived. Table 11 summarizes the main points of the newly proposed data-driven fleet prognostics type definition in comparison with the original conceptual definition provided by Al-Dahidi et al. (2016). While, in general, the proposed concept has been retained, adjustments in terms of more detailed elaborations and clear differentiations have been introduced, taking into account the data-driven trajectory behavior analysis. The adjusted description clearly distinguishes type I from type II and type III, whereas the latter two types differ only considering the data-driven perspective. For homogeneous and heterogeneous fleets, the influencing factors may be similar; however, their effect on the degradation trajectories is different. For prognostics, this differentiation, though, is of major importance. The extended fleet definition can therefore be used to conceptually distinguish the identical fleet from the homogeneous and heterogeneous fleet, while the classification of homogeneous and heterogeneous fleets must be performed with a data-driven perspective.

Table 11 | Summary Fleet Prognostic Type Definition

	Identical Fleet Type I	Homogeneous Fleet Type II	Heterogeneous Fleet Type III
Original Conceptual Definition	have identical features and usage and work in the same operating conditions	share some technical features and work in similar operating conditions, but show differences either on some features or on their usage	have different and/ or similar technical features but undergo different usages with different operating conditions
Adjusted Description	have machine-to-machine variations (same configuration), but with identical usage and operated in the same environment	have machine-to-machine variations and are influenced by one or multiple external factors and/ or different machine configurations	have machine-to-machine variations and are influenced by one or multiple external factors and/ or different machine configurations
Data-Driven Perspective	identical degradation behavior with variations in terms of random or systematic noise between machines	varying degradation behavior influenced by external factors with no distinct sub-fleets	different degradation behavior influenced by external factors with distinct sub-fleets

With regard to the resulting requirements for fleet prognostics, it can be concluded that approaches for identical fleets need to handle machine-to-machine variations reflected by random or systematic noise. In contrast, approaches for homogeneous and heterogeneous fleets must additionally address further influencing factors. If available and attributable to a certain behavior, the influencing factor should be included as a covariate of the algorithm development. If this is not the case, the dissimilar behavior should be accounted for by the algorithm itself. In addition, as the heterogeneous fleet depicts clearly separated subgroups with similar behavior, these sub-groups should be targeted individually, or the approach should be able to learn and distinguish the sub-fleets.[18]

4.3.4 Fleet Prognostic Type Characterization

With respect to the identified fleet prognostic types, the similarity of data, as well as the existence of homogeneous subsets, needs to be investigated in order to derive the fleet types by data analysis. The *similarity assessment of fleet data* should focus on the degradation behavior of the fleet members, as this information is the main constituent for generating a remaining useful life estimation. For this purpose, similarity measures offer a good and commonly applied possibility. These measures calculate the distance between two trajectories and therefore provide insights on the structure and composition of the fleet with regard to prognostics. Similarity measures are already used in the field of PHM with fleet data. Applications include sub-fleet identification based on the highest degradation pattern similarity (e.g., Leone et al. 2017) and similarity-based prognostic approaches (e.g., Wang 2010; Bektas et al. 2019).

Furthermore, the *existence of homogeneous subsets* within the fleet has to be investigated to improve the distinction between the homogeneous fleet and the heterogeneous fleet. This can be done via clustering. In general, clustering targets the discovery of similar groups of objects with comparable characteristics (Kammoun and Rezg 2018, p. 1094). It is a descriptive task that aims to identify groups with high intra-group homogeneity and inter-group separability (Rokach and Maimon 2005, p. 321). In contrast to classification, it is an unsupervised approach with unlabeled data. Though, it provides potential data labels by identifying hidden traits from a heterogeneous data set (Kammoun and Rezg 2018, p. 322). Fleet clustering is a concept used in diagnostics. To identify similar machines, clusters are built using working conditions and configuration parameters (Lee et al. 2014a, p. 6; Lapira 2012, p. 6). Another approach is fault detection based on pattern similarity. Using hierarchical clustering, machines in the same condition (normal vs. faulty) exhibit shorter distances and are thus clustered (Hendrickx et al. 2019, p. 104).

In order to identify homogeneous subsets given degradation data from a fleet of machines, approaches provided by *time-series clustering* can be applied. Time-series clustering is a specific case of clustering, which handles dynamic data instead of static objects (Liao 2005, p. 1858). Time-series are specific temporal sequences with continuous and real-valued elements (Antunes

---

[18] These requirements will be taken up, detailed, and further elaborated in Chapter 5.

and Oliveira 2001, p. 2; Aghabozorgi et al. 2015, p. 17). They can have different characteristics, ranging from univariate or multivariate data, uniformly or non-uniformly sampled data, as well as equal or unequal sequence length (Liao 2005, p. 1858). Data obtained through machine condition monitoring, therefore, generally depicts a time-series and thus, allows for the use of time-series clustering.

There are two ways to handle time-series data for clustering. On the one hand, the clustering algorithm can be tailored to time-series data by generally adjusting the distance measure. This approach is denoted as raw data-based approach (Liao 2005, p. 1859) or shape-based approach (Aghabozorgi et al. 2015, p. 19). It requires data to be sampled at the same interval; however, it is able to deal with unequal sequence length (Liao 2005, p. 1864). Another option is the transformation of dynamic time-series data to static objects (Aghabozorgi et al. 2015, p. 19). Here, feature-based approaches reduce the dimensionality of the data set to a few informative features, while model-based approaches fit models to raw data and use the obtained parameters for clustering (Rani and Sikka 2012, p. 1; Liao 2005, p. 1859). For the assignment of fleet prognostic types, the raw data-based approach is preferable since no further assumptions or specifications are to be made. In contrast, feature-based approaches require the definition of features, which are highly application dependent and do not allow for generalization. The same applies to model-based approaches. In addition, model-based approaches scale poorly and have bad performance when clusters are closely adjacent (Aghabozorgi et al. 2015, p. 20).

Due to the dynamic nature of time-series, some specific aspects must be considered in contrast to the clustering of static objects. These include time-series representation, similarity and distance measures, clustering prototypes, and time-series clustering (Aghabozorgi et al. 2015, p. 20). Taking into account these aspects, a process for the data-driven characterization of fleet data can be elaborated. For this purpose, the clustering pipeline described by Adolfsson et al. (2019, p. 15), as well as the CRISP-DM methodology for data analytics (cf. Chapter 3.3.2), is taken as a basis and adapted to time-series clustering as well as to fleet prognostic type identification.[19] The resulting process, as well as the details regarding the concrete realization of each process step, will be presented in the following chapter as part of the characterization method for fleet prognostics.

4.4 Characterization Method Design and Construction

Drawing on the previous results, the fleet prognostic characterization method is presented in this chapter. It highlights relevant steps to describe the fleet and analyzes its data to prepare its usage within fleet prognostics. In contrast to existing literature, the approach targets a data-

---

[19] For this purpose, the detailed descriptions of time-series clustering by LIAO (2005) , AGHABOZORGI ET AL. (2015) as well as SARDÁ-ESPINOSA (2017) have been respected.

driven perspective to better ambiguity, uncertainty, and lack of knowledge about the specific behavior of the fleet. Figure 30 presents the elaborated fleet prognostics characterization process. First, a general understanding of the fleet and its context is targeted and structured, given the available expert knowledge as well as the fleet data (cf. Chapter 4.4.1). This is followed by a deeper investigation of the fleet data to identify the corresponding fleet prognostic type, which demonstrates the fleet data characteristics and the machine's heterogeneity with regard to their degradation behavior. For this purpose, a data-driven process for the identification of the fleet prognostic types is presented (Chapter 4.4.2), and details are provided for each of the steps (Chapters 4.4.3, 4.4.4, and 4.4.5). The fleet description, as well as the fleet prognostic type, are required to better accommodate the fleet-specific characteristics during fleet prognostics (which will be discussed in Chapter 5).

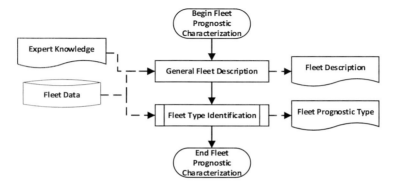

Figure 30 | Fleet Prognostics Characterization Process

4.4.1 General Fleet Description

The first part of the characterization method for fleet prognostics depicts a conceptual analysis of the fleet in order to gain a deeper understanding and provide the foundation for the data-driven fleet data analysis.[20] For this purpose, the fleet needs to be specified with regard to the four dimensions identified. In addition, the detailed categories, as well as the specific characteristics, are outlined. The description of the fleet should be carried out from the point of view of the complete fleet. In case there are different characteristics for one fleet, all characteristics should be specified.

Application Area | The first dimension for fleet characterization depicts the application context in terms of the unit of analysis as well as the level of implementation of the fleet and the addressed PHM task:

---

[20] The general fleet description is based on the information gathered from the first round of the literature review, and further extended by the information gathered within the second round.

- Unit of Analysis: e.g., Automotive Vehicle | Locomotive | Ship | Industrial Robot
- Level of Implementation: System | Sub-System| Component | Equipment
- PHM Context: Fault Detection | Diagnostics | Prognostics

Fleet Composition | The fleet composition dimension reflects the variety of the machines in a fleet and their main disparity. In particular, the dimension addresses the nature of heterogeneity of a fleet with respect to different technical and functional characteristics of machines within a fleet:

- Size: Number of Machines
- System Design: Identical | Variations | Variants | Customized
- Nature of Heterogeneity: Mechanical | Electrical | Electronic | Software
- Age Structure: Identical | Different | Dynamic
- Manufacturer: Same | Different
- Functioning: Identical | Similar | Non-Identical
- Failure Modes: Single | Multiple
- Position: Identical | Different

Operating Condition | The operating condition refers to the operational parameters the fleet is exposed to. Even though the operating condition is often referred to as one concept, it comprises the usage (time), the load/ stress, as well as the environmental conditions (with regard to climatic conditions and the physical environment (Ghodrati 2005, p. 55)). In order to specify the operating conditions of the fleet, the following characteristics can be used:

- Usage: Identical (Stationary) | Cluster (fixed[21]) | Cluster (changing[21]) | Individual
- Load/ Stress: Identical (Stationary) | Cluster (fixed[21]) | Cluster (changing[21]) | Individual
- Environment: Identical (Stationary) | Cluster (fixed[21]) | Cluster (changing[21]) | Individual

Data | The data dimension describes and analyzes the available data for fleet PHM. In general, three areas for data analysis can be distinguished, namely, data description, data properties, and data transmission. Data description refers to the structural description of the data values, while data properties specify the contextual information of the data. Lastly, data transmission determines when data becomes available for analysis. The different characteristics of the data dimension are as follows:

Data Description
- Structure: Unstructured | Semi-structured | Structured
- Values: Continuous | Discrete | Textual

---

[21] The characteristic *cluster* refers to the operating condition of the fleet, which can be classified into distinct clusters. Within fixed clusters, machines are assigned to a specific cluster, while in the case of changing clusters, the operating conditions of machines can change dynamically within the specified cluster.

- Dimension: Single | Multiple
- Stationarity: Stationary | Non-Stationary
- Types: Raw Signals | Process Information

Data Properties
- Generation: Simulation | Experiment | Real
- Run-to-Failure: Incomplete | Complete
- Acquisition Time: Cycle | Time-varying

Data Transmission
- Transmission: Real-time | Online | Offline

The four different dimensions, application area, fleet composition, operating condition, and data, are identified to describe and categorize fleets in the context of PHM. Figure 31 summarizes the identified characteristics of the general fleet description. The first dimension, the application area, describes the addressed application context. For a fleet to be useful for prognostics, the unit of analysis, the level of implementation as well as the application context should be equal for all machines in the fleet, as those machines share at least some basic characteristics. The same application area of all machines in the fleet is, therefore, an essential prerequisite for further usage within the fleet prognostics. The two dimensions of fleet composition and operating condition conceptually describe the heterogeneity between machines in a fleet. The benefits of fleet data for prognostics are greater the more similarities the machines within the fleet have in terms of their design and operating conditions. While identical system design, usage, load, and environment will result in similar degradation patterns, customized design, highly individual usage, and different environmental conditions can strongly influence the deterioration behavior, reducing or even negating the ability to use the fleet data for prognostics. Lastly, the data dimension represents the characterization of the available data. This information is crucial for general prognostic algorithm development; however, they are largely independent of the fleet. Notwithstanding, as important information is provided, which should be considered within prognostics, this dimension qualifies to be included in the general fleet description.

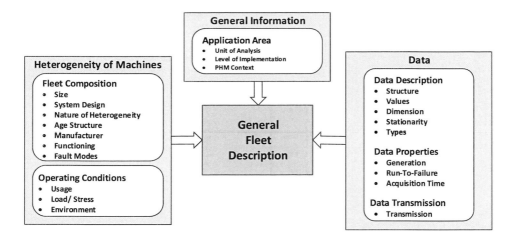

Figure 31 | Overview of the General Fleet Description

4.4.2 Fleet Prognostics Type Identification Process

Following the conceptual fleet description, an in-depth analysis of the available fleet data is required to identify the fleet prognostic type and by thus, define requirements for fleet prognostics. For this purpose, an approach is required that is able to systematically analyze the available fleet data in this regard. While available influencing factors or expert knowledge can provide first insights on the fleet structure, a detailed analysis of the data is essential to obtain a better understanding of the composition of the fleet from a data-driven perspective. In the presence of additional knowledge regarding differences in the fleet, this information can be used subsequently to validate the resulting fleet prognostic type and assign the distinct group labels. The developed process mainly focuses on the distinction between the homogeneous and heterogeneous fleet as the identical fleet is already easily deducible from its context (same machine configuration with identical usage and environment) as well as through visual degradation assessment. The division into the homogeneous and heterogeneous fleet, however, is uncertain given further influencing factors and thus needs to be identified considering the different degradation trajectories. This is targeted in the fleet prognostic type identification process.

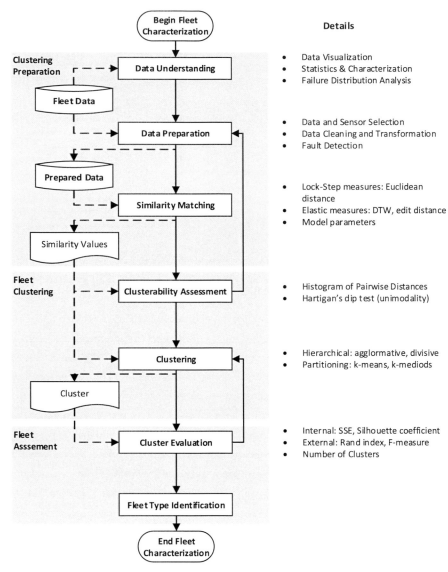

Figure 32 | Fleet Prognostic Type Characterization Process

The fleet prognostic type identification process comprises the phases of *clustering preparation, fleet clustering,* and *fleet assessment.* Analysis should, therefore, start with a simple data understanding step to gain a first impression of the fleet data at hand. This is followed by general data preparation to convert raw data into cleaned data. Given these prepared data, a similarity matrix can be calculated. This similarity matrix is input to the fleet clustering phase. Prior to clustering, clusterability is examined to identify if the data exhibit a certain clustering structure. This is followed by clustering. Lastly, resulting clusters are evaluated on their quality, and the fleet prognostic type can be defined. The complete process is depicted in Figure 32, including a short overview of possible configurations of each step on the right side of the figure, as identified by the literature study of time-series clustering and clustering in general. The individual phases and their steps, as well as the design options, are described in detail in the following chapters.

4.4.3 Clustering Preparation Phase

The phase clustering preparation includes the steps required for the execution of clustering algorithms. After gaining the first insights into the available data, data needs to be reduced, cleaned, and transformed. Lastly, a similarity assessment between two machines has to be specified and the similarities calculated.

Data Understanding | Before starting with data preparation and the clustering analysis, the first impression of the data needs to be obtained. Carrying out simple statistics and plotting the data supports this step. In addition, data can be categorized based on its behavior (e.g., ascending, binary) to provide an overview and support data selection afterward. Furthermore, the time-to-failure probability distribution of all machines within the fleet can give the first indications on multimodality, disregarding the actual degradation behavior.

Data Preparation | Data need to be prepared adequately in order to gain substantial and reliable insights into the fleet; otherwise, the cluster results may be distorted by, e.g., noise. Preparing data for time-series clustering or prognostics can, to a large extent, be carried out in a similar manner. Nevertheless, the data preparation step is, of course, highly dependent on the underlying data set and can therefore only be standardized to a certain degree. Various steps for data preparation in fleet prognostics have been identified through the structured literature review and will be presented in detail in Chapter 5.4.2.1. Given these results, a subset can be derived for data preparation in the context of time-series clustering. This subset represents the generic tasks that may be required, with describing specific methods of implementation in detail, as this is beyond the scope of this chapter. Therefore, the selection of the specific method for a specific task should take into account both the characteristics of the available methods and their compatibility with the data set, as well as the capabilities of the user.

The key tasks to be performed for data preparation in the fleet prognostic type identification process include:

Data and Sensor Selection: Data sets often contain numerous attributes. These have to be reduced to a practical and meaningful subset. Both expert knowledge and data analysis and visualization can be employed. Furthermore, the prognostics-specific methods are applicable (as described in 5.4.2.1), as the same data subset will most likely be used for the implementation of the prognostic algorithm.

Data Cleaning and Transformation: Data cleaning comprises various procedures to repair data and enhance its quality. Furthermore, to adjust data to the same scale and clean the influence of dynamic working conditions, normalization and respectively multi-regime normalization (described in 5.4.2.1) are advisable. In addition, to reduce noise, the time-series should be smoothed adequately. Popular methods in this context include exponential curve fitting, moving average filtering and interpolation, and kernel regression smoothing (Wang 2010, pp. 42–45).

Fault Detection: To identify the period of the main degradation, fault detection can be applied. Here, initial faulty behavior is identified by an algorithm, and further analysis no longer takes into account the part of the time-series under normal behavior. Given the reduced time-series, a plot of the time-to-failure probability distribution of the fleet can be examined to identify potential clusters with respect to the failure distribution, independently of the shape of the time-series.

Further steps, such as feature extraction, health state estimation, or data reduction, should be carried out only for prognostics, as crucial information could be lost or homogenized during transformation. In addition, further steps may certainly be necessary depending on the specific application. In general, data should be transformed into a format that allows a comparison of the degradation behavior between two time-series.

Similarity Matching | The similarity between instances must be defined to identify clusters with comparable candidates. There are various concepts to specify similarity and dissimilarity (Santini and Jain 1999, p. 2). For time-series assessment and clustering, distance measures are applied. The distance between two time-series F_i and F_j is calculated by the sum of the distances between the measures f_i and f_j of the two time-series over the duration of the time-series T (Aghabozorgi et al. 2015, p. 23):

$$dist(F_i, F_j) = \sum_{t=1}^{T} dist(f_{it}, f_{jt}) \tag{4.1}$$

The computed distances between all n instances result in a nxn-distance matrix, also referred to as a proximity matrix (Kaufman and Rousseeuw 2005, p. 20), which can be used subsequently for clustering. As opposed to traditional distance calculation, distances of time-series are often approximations due to different lengths and sample intervals (Aghabozorgi et al. 2015, p. 23). Further challenges include noise, amplitude scaling, offset translation, longitudinal scaling, linear drift, discontinuities, and temporal drifts (Aghabozorgi et al. 2015, p. 23).

Numerous distance measures are available for time-series dissimilarity analysis (Wang et al. 2013; Aghabozorgi et al. 2015, p. 23; Liao 2005, pp. 1861–1863), which can be structured based on three objectives, namely similarity in time, shape, and change (Aghabozorgi et al. 2015, p. 24).

Distance measures that seek similarity in time are *lock-steps measures*, which perform a direct (one-to-one point) analysis of the values of the time-series (Wang et al. 2013, p. 280). The most common metric is the Euclidean distance (Sardá-Espinosa 2017, p. 4; Aghabozorgi et al. 2015, p. 23) as well as other L_p-norm measures (Wang et al. 2013, p. 280; Sardá-Espinosa 2017, p. 4). Their greatest strength is the simplicity, linear complexity, and the lack of parameters (Wang et al. 2013, p. 281). On the other side, these measures are noise sensitive and cannot handle scale and time shifts (Sardá-Espinosa 2017, p. 4). Due to their one-to-one time point analysis, lock-step measures are only applicable for time-series with equal length (Sardá-Espinosa 2017, p. 4). For the case of fleet identification, the time-series under investigation needs to be adjusted to have an equal size. Thus, to perform distance analysis of time-series with unequal length by finding the similarity in shape, *elastic measures* are applicable, which implement a one-to-many mapping and therefore cope with several restrictions of lock-step measures (Wang et al. 2013, p. 282; Sardá-Espinosa 2017, p. 5). The most known member of elastic measures is dynamic time warping (DTW) (Aghabozorgi et al. 2015, p. 25). DTW searches for a non-linear alignment for two time-series such that the difference is minimized (Liao 2005, p. 1862). This alignment is referred to as a warping path. A lock-step measure can then be used to calculate the distance of the warping path. To reduce the computational effort, a time window can be defined for the identification of the best path (Sardá-Espinosa 2017, p. 7). Other elastic measures are based on the edit distance, as for example, the longest common subsequence (LCSS) (Aghabozorgi et al. 2015, p. 25) and the edit distance on real sequences (EDR) (Wang et al. 2013, p. 277). Edit distance defines the distance based on the number of operations required to transfer one time-series to the other one. For this, a threshold of similarity is specified, which defines a match (Wang et al. 2013, p. 282). In comparison to DTW, edit distance measures permit non-matching time points (Wang et al. 2013, p. 282). Lastly, while similarity in time and shape targets short-length time-series, the similarity in change focuses more on the global structural level (Aghabozorgi et al. 2015, p. 24). For this purpose, models are developed to describe the time-series, and a similarity assessment is performed considering the values of the parameters (Aghabozorgi et al. 2015, p. 24). In this context, the parameters of the fitted models for smoothing could be used as well as an additional simple distance measure.

The presented measures depict an excerpt of the most frequently mentioned distance measures for time-series analysis. An in-depth study on various distance measures, however, revealed that the well-established measures Euclidean distance, DTW, and edit distance generally score well on different data sets, while novel measures are often inferior (Wang et al. 2013, p. 305). As the selection of the distance measure has a great impact on the result (Liao 2005), other measures should be considered if the results are of inferior quality (Wang et al. 2013, p. 305). In order to support the selection of an appropriate measure for a given application, Table 12 summarizes the main advantages and disadvantages of the four mentioned well-known distance measures.

Table 12 | Distance Measures

Distance Measure	Advantages	Disadvantages
Euclidean Distance (EUC)	• Simplicity and easy to implement (Wang et al. 2013, p. 281) • Linear complexity (Wang et al. 2013, p. 281) • Parameter-free (Wang et al. 2013, p. 281)	• Lock-Step Measure (requires truncation) (Aghabozorgi et al. 2015, p. 22; Sardá-Espinosa 2017, p. 4) • Not able to handle local temporal drifts (Wang et al. 2013, p. 281) • Sensitive to scaling (Aghabozorgi et al. 2015, p. 22), noise, and misalignments in time (Wang et al. 2013, p. 281)
Dynamic Time-Warping (DTW)	• Elastic measure (Aghabozorgi et al. 2015, p. 22) • Able to handle temporal drift (Aghabozorgi et al. 2015, p. 22) • Parameter-free (Wang et al. 2013, p. 282)	• Computational demanding/ low efficiency (Aghabozorgi et al. 2015, p. 22; Sardá-Espinosa 2017, p. 5) • Sensitive to noise (Chen et al. 2005, p. 2) • No metric (does not follow triangle inequity) (Chen et al. 2005, p. 2)
Longest-Common Subsequence (LCSS)	• Robust (Vlachos et al. 2002, p. 1) • Elastic measure (Aghabozorgi et al. 2015, p. 26) • Unequal size time-series (Aghabozorgi et al. 2015, p. 26)	• Requires parameter tuning • Polynomial-time complexity (Chen et al. 2005, p. 4) • No metric (does not follow triangle inequity) (Chen et al. 2005, p. 2)
Edit Distance on Real Sequence (EDR)	• Handles local time-shifting (Chen et al. 2005, p. 2) • Effects of outliers are reduced (Chen et al. 2005, p. 3) • Matching threshold (Chen et al. 2005, p. 3) • Elastic measure (Aghabozorgi et al. 2015, p. 22)	• Requires parameter tuning • Polynomial-time complexity (Chen et al. 2005, p. 4)

4.4.4 Fleet Clustering

The second phase takes the computed nxn-distance matrix as input to perform clustering. To draw a conclusion about the cluster quality and existing data structure, primarily clusterability assessment is performed. This is followed by clustering.

Clusterability Assessment | The analysis of clusterability should generally precede clustering (Adolfsson et al. 2019). To obtain informative clusters, it must be determined whether the data exhibits a proper cluster structure. Only in the case of a non-random structure; clustering analysis will return appropriate results; otherwise, the clusters may be both deceptive and insignificant (Adolfsson et al. 2019, p. 2; Han et al. 2012, p. 484). To assess clusterability, different statistical tests are available. Statistical tests of multimodality evaluate the presence of multiple modes in data. There are two statistical tests for this purpose, namely the dip test and the Silverman test (Adolfsson et al. 2019, p. 6). While the former makes use of distance measures, the latter requires principal curves or components as data reduction (Adolfsson et al. 2019, p. 6). Apart from that, the Hopkins statistic tests the uniformity of the data distribution (Han et al. 2012, pp. 484 f.). In general, only if these tests are not rejected, data entails a clusterable structure, and clustering will result in meaningful results (Adolfsson et al. 2019, p. 2).

For the assessment of clusterability in the context of fleet characterization, Hartigan's dip test for unimodality is proposed due to the use of pairwise distances. Hartigan's dip test analyzes data assuming unimodal data distribution as the null hypothesis. Multimodality, on the other hand, is defined as the alternative hypothesis. As the test, however, is designed for continuous data, it is not applicable for distance measures yielding discrete values (e.g., based on edit distance). Therefore it is additionally recommended to visualize pairwise distances in a histogram prior to testing. This visualization can be used as the first indicator of multimodality, which is the case if there are two distinct groups visible with small and larger distances (Adolfsson et al. 2019, p. 6).

Figure 33 illustrates the histogram of pairwise distances using DTW as a distance measure as well as the corresponding p-values provided by Hartigan's dip test with regard to the four simulated fleet data introduced in Chapter 0. These four data sets have been simulated, taking into account the derived fleet data characteristics. For the identical and homogeneous fleets, the distance values between the machines are clearly distributed unimodally, yielding a p-value > 0.95 (confirming the null hypothesis with no prevalent clustering structure). On the contrary, both heterogeneous fleets exhibit multimodality with regard to the pairwise distance values with p-value < 0.05.

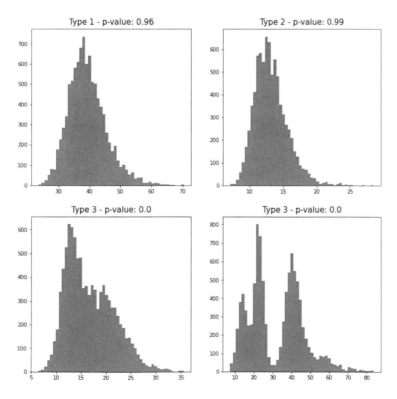

Figure 33 | Clusterability Evaluation – Distance Histogram

Clustering | In case the clusterability assessment reveals a possible cluster structure in the data, clustering can be performed. For this, the degradation trajectories can be grouped to form similar sub-fleets, and further analysis should focus only on the identified sub-fleets. To do so, a huge amount of different clustering methods are applicable, which can be summarized in five groups: hierarchical, partitioning, model-based, grid-based, and density-based methods (Han et al. 2012, pp. 448–450; Liao 2005, p. 1858). While clustering methods in the first two groups are frequently used in the context of time-series analysis, model-based, grid-based, and density-based methods receive little recognition (Aghabozorgi et al. 2015, pp. 27 f.; Rani and Sikka 2012, p. 1; Sardá-Espinosa 2017, p. 2). For this reason, in the following, hierarchical and partitioning methods are shortly explained, and their well-known methods are listed briefly.

Hierarchical methods. This group iteratively partitions the data into clusters. Partitioning can either be done *agglomerative* (bottom-up) or *divisive* (top-down). In order to combine or split clusters, distance measures between two clusters are defined. Most known is the single-linkage (minimum distance between two members of different clusters), complete-linkage (maximum distance), and average-linkage (average distance) clustering approaches (Rokach and Maimon 2005, p. 331). Results of hierarchical methods are presented using dendrograms. This result visualization enables the user to select an appropriate number of clusters without having to specify it in advance (Rokach and Maimon 2005, p. 331; Aghabozorgi et al. 2015, p. 33). Nevertheless, hierarchical methods have a slow performance for bigger data sets and do not allow for an adjustment of previous cluster consolidations or separations (Liao 2005, p. 1858; Rokach and Maimon 2005, p. 332).

Partitioning methods. Building on a starting clustering solution, partitioning methods heuristically investigate different decompositions of the data and assign them to a predefined number of clusters (Rokach and Maimon 2005, p. 332). The *k-means* method is the most applied partitioning method (Aghabozorgi et al. 2015, p. 26). Through the iterative realization of two steps ((1) adaptation of cluster allocation and (2) updating of cluster centers), an objective function which calculates the distance to the cluster centers is minimized (Liao 2005, p. 1860). For *k*-means, the cluster center (prototype) is defined by the mean value of its members. This is straightforward for time-series of equal length; nevertheless, for unequal sequence length, the mapping is complex (Aghabozorgi et al. 2015, p. 28). In contrast, its variant *k-medoids* assigns its most centric object as cluster representative, which is defined as the sequence that minimizes the sum of squared distances to all other instances in the cluster (Aghabozorgi et al. 2015, p. 28). Further variants include fuzzy c-means and fuzzy c-medoids. Here, instead of defining unique cluster assignments, the likelihood of different cluster memberships is calculated (Liao 2005, pp. 1860 f.). Partitioning methods are characterized by their fast processing time and their high quality (Aghabozorgi et al. 2015, p. 33; Kaufman and Rousseeuw 2005, pp. 44 f.). In contrary to hierarchical methods, however, the number of clusters needs to be defined beforehand, which makes them difficult to apply if no prior information is available (Rokach and Maimon 2005, p. 332).

Due to its descriptive role, multiple methods are often explored, and resulting clusters are compared (Rokach and Maimon 2005, p. 321). Criteria for the selection of an appropriate method include, among others, their computational time complexity, the need for expert knowledge (in terms of the number of resulting clusters) as well as their performance. However, a detailed study has shown that the choice of method has no significant influence on the results (Liao 2005, p. 1871). As a result of this phase, a proposal for a cluster structure of the machines into more homogeneous sub-fleets is provided. This partitioning has to be assessed and evaluated in the subsequent phase.

4.4.5 Fleet Assessment

The last phase, fleet assessment, takes as input the identified cluster and evaluates their quality considering external and internal indices. In the last step, fleet prognostic type definition, the fleet prognostic type is specified given the obtained results.

Cluster Evaluation | In order to draw a conclusion on possible fleet clusters, the identified clusters need to be evaluated. In general, this is done using standardized performance metrics referred to as cluster validity indices (Sardá-Espinosa 2017, p. 23). Evaluation indices are separated into the groups internal and external indices (Rokach and Maimon 2005, p. 326).

External indices evaluate the clustering taking into account the ground truth. While clustering is an unsupervised method, labels might be available due to synthetic data or using human judges (Aghabozorgi et al. 2015, pp. 30 f.). Two well-established and commonly applied indices in the context of time-series clustering are the Rand index and the F-measure (Rokach and Maimon 2005, pp. 329 f.; Aghabozorgi et al. 2015, p. 32). The *Rand index* calculates for each pair of instances in a data set whether the ground truth matches the clustering. For this purpose, the number of matches between the ground truth and the clustering (a denotes the same cluster and b different clusters) as well as the number of mismatches (denoted by c and d) are determined. The Rand index is then defined as follows (Rokach and Maimon 2005, p. 330):

$$R = \frac{a + b}{a + b + c + d} \tag{4.2}$$

The *F-measure* is a cluster validity index, which is based on recall (accuracy of the actual positive values) and precision (accuracy of the predicted positive values) (Rokach and Maimon 2005, pp. 329 f.). The F-measure is calculated as:

$$F = 2 * \frac{Precision * Recall}{Precision + Recall} \tag{4.3}$$

Clustering in real-world applications is generally purely descriptive, and no labels are available (Aghabozorgi et al. 2015, p. 30). For this purpose, *internal indices* assess the quality of the cluster structure in terms of cluster separations (Aghabozorgi et al. 2015, p. 32; Han et al. 2012, p. 489). The objective of clustering is a high intra-cluster similarity and inter-cluster distance (Aghabozorgi et al. 2015, p. 32). The *sum of squared error* (SSE) index is a simple and frequently used index for internal evaluation (Aghabozorgi et al. 2015, p. 32; Rokach and Maimon 2005, p. 326). It is defined as follows:

$$SSE = \sum_{k=1}^{K} \sum_{\forall x_i \in C_k} \|x_i - \mu_i\|$$

(4.4)

where C_k depicts all instances in cluster k and μ_i is the mean value with respect to cluster k (Rokach and Maimon 2005, p. 326). Besides the sum of squared error-index, the *Silhouette coefficient* is well-known for internal cluster analysis. For each instance, the Silhouette is calculated using the average distance of the instance to all other instances in the same cluster (a_i) as well as the average distance to all objects in the nearest neighboring cluster (b_i) (Kaufman and Rousseeuw 2005, pp. 83–85; Rousseeuw 1987, pp. 55 f.):

$$SH_i = \frac{b_i - a_i}{\max(a_i, b_i)}$$

(4.5)

The Silhouette is defined in the interval [-1, 1] (Kaufman and Rousseeuw 2005, p. 85). The Silhouette plot provides a simple visualization of all Silhouettes and can therefore well represent the quality of the assignment of the individual instances. The Silhouette coefficient then denotes the average Silhouette value for the complete data set (Kaufman and Rousseeuw 2005, p. 87):

$$SC = \frac{1}{|C|} \sum_{i \in C} SH_i$$

(4.6)

This coefficient provides a good indicator of the overall clustering quality. In general, $SC > 0.71$ implicates a strong clustering structure, whereas SC < 0.25 implies the lack of a substantial clustering structure (Kaufman and Rousseeuw 2005, p. 88).

The *number of clusters k*, which is required for partitioning methods, can be determined analytically. While there are numerous approaches to identify the number k (Kodinariya and Makwana 2013, p. 90), two approaches are explained in detail, which make use of cluster validity indices presented above. The most simple and commonly applied approach is the *Elbow method*. Here, the average within-cluster distance for various numbers of clusters is visualized by a line chart (Thorndike 1953, p. 275). The suitable number of clusters is then defined as the value of k where the decrease in the within-cluster distance measures stagnates (Kodinariya and Makwana

2013, p. 92). In addition, the Silhouette score can be used for determining the number of clusters. In this case, the average Silhouette score is calculated for different values of k. The best number of clusters is then defined as k, which maximizes the average Silhouette score (Kaufman and Rousseeuw 2005, p. 87). In the case of fleet prognostic type identification, $k > 1$ strongly indicates the existence of a heterogeneous fleet, where k representing the number of homogeneous sub-fleets within the heterogeneous fleet.

Fleet Prognostic Type Identification | Results from the clusterability and cluster evaluation can be used to assign the specific fleet prognostic type to the data under investigation. External evaluation indices assess the clustering quality regarding the ground truth; thus, they do not provide any further information on the fleet structure other than validating the cluster assignment. Internal indices, on the other side, analyze the inter-cluster homogeneity and intra-cluster separability and can therefore be used for fleet prognostic type identification. While the sum of the squared error index highly depends on the data values, meaning that the calculated value can only be seen in comparison with other clustering results, the Silhouette coefficient provides a range-independent standardized assessment. It can therefore be taken for fleet prognostic type characterization. With respect to the Silhouette coefficient interpretation by Kaufman and Rousseeuw (2005, p. 88), the following conclusions can be drawn:

- $SC \leq 0.25$: No substantial clustering structure has been identified
- $0.25 < SC \leq 0.50$: The clustering is weak (might be artificial)
- $0.50 < SC \leq 0.70$: The clustering structure is reasonable
- $SC > 0.70$: The clustering structure is strong

These data-driven insights with respect to the fleet data under investigation results in the following fleet distinction:

Identical Fleet: The identical fleet, which is described by the same machine configuration with the same usage and environment, has been defined as part of the fleet description step. To validate the prior fleet assignment, the presented fleet prognostic type identification process can be applied. For the identical fleet, the degradation behavior is assumed to be equal among all machines yielding high similarity between machines (machine-to-machine heterogeneity with minor machine deviation due to noise) with no cluster structures. As a result, the identical fleet should exhibit a high p-value $(p > 0.05)$ during clusterability assessment as well as a small Silhouette coefficient (SC $\leq$ 0.25).

Homogeneous Fleet: The degradation trajectory of homogeneous fleets is characterized by influencing factors. As the homogeneous fleet does not exhibit any strong sub-fleet structure, the fleet prognostic type characterization process should not discover clusterable structures. For this reason, the homogeneous fleet is characterized by a p-value above 0.05 and a Silhouette coefficient below 0.25, matching the values from the identical fleet.

Heterogeneous Fleet: Like the homogeneous fleet, the heterogeneous fleet is also shaped by influencing factors. However, there are clearly distinguishable sub-fleets. The process, therefore, assigns the heterogeneous fleet for data with a small p-value (below 0.05) and a high Silhouette coefficient (above 0.7), which indicates a strong clustering structure.

The presented assignments are cases with a very strong indication of the underlying fleet, but they do not comprise the interval between $0.25 < SC \leq 0.70$. For this range, only a weaker assignment is possible. In general, it can be said that the smaller the SC value, the more reasonable the assignment of the homogeneous fleet and vice versa. In accordance with the above presented SC interpretation, a weaker fleet prognostic type definition is proposed, with an $SC \leq 0.50$ representing a homogeneous fleet and $SC > 0.5$ a heterogeneous fleet. Besides these soft constraints, the p-value should be considered as well as a visual inspection of the fleet degradation behavior is suggested. Table 13 provides an overview of the fleet prognostic type assignment with respect to the p-value and the Silhouette coefficient.

Table 13 | Fleet Prognostic Type Identification

Fleet Prognostic Type	Purpose	Strong Assignment		Weak Assignment
		p-Value	Silhouette Coefficient	
Identical Fleet	Validation	p-Value >= 0.05	SC <= 0.25	
Homogeneous Fleet	Assignment	p-Value >= 0.05	SC <= 0.25	SC <= 0.50
Heterogeneous Fleet	Assignment	p-Value < 0.05	SC > 0.70	SC > 0.50

Figure 34 shows the Silhouette coefficient for the four simulated fleets. It can be seen that both the identical fleet (plot: upper left) and the homogeneous fleet (plot: upper right) depict an average Silhouette below 0.25. For the heterogeneous fleet, however, the picture is not entirely clear. While both fleets show heterogeneous behavior with respect to the p-value as well as through visual inspection, the assignment is not definite considering the Silhouette coefficient. For the heterogeneous fleet with the same functional behavior (plot: lower left), the average Silhouette coefficient is below 0.5 and could even indicate a homogeneous fleet. However, as the p-value suggests a prevalent cluster structure, which is confirmed by visual inspection, the heterogeneous fleet is assigned. Nevertheless, for example, in the case of a few machines in a fleet, the homogeneous fleet can be assigned depending on the application scenario and expert knowledge. For the heterogeneous fleet with different functional degradation behavior (plot: lower right), the average Silhouette is above 0.5 and thus indicates a heterogeneous fleet with a weak assignment. The heterogeneous fleet is confirmed with respect to the p-value as well as by visual examination. Given these four fleets, in general, it can be seen that the Silhouette coefficient for time-series is rather small, even for the two cases with clear clustering structures. This is attributed to the fact that the distance between the time-series is quite small due to similar behavior (increasing trend), even between different classes. With respect to the distance measure, the intra-group homogeneity is, therefore, only slightly lower than the inter-group separability, which is reflected by the Silhouette coefficient. For this reason, fleets that are subject to weak assignments should be carefully inspected.

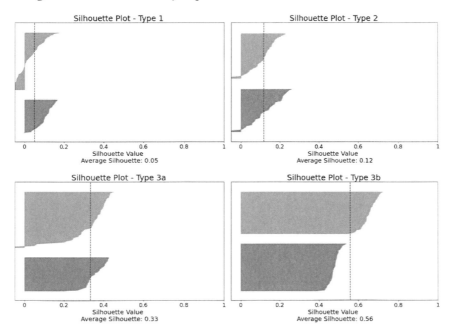

Figure 34 | Fleet Prognostic Type Identification – Silhouette Coefficients (Example)

The fleet description presented in Chapter 4.4.1 can be used to validate the identified fleet prognostic type and provides additional information about the underlying reason for the assignment. In particular, the information gathered from the dimensions of fleet composition and operational condition can be leveraged for this purpose. In the case of the heterogeneous fleet assignment, the fleet description enables the identification of possible factors causing the heterogeneity of the fleet. Here, it can be examined whether the characteristics of the machines within the fleet correspond to the identified cluster. An example would be a heterogeneous fleet with four identified clusters ($k = 4$) and a customized system design. In this case, it would be possible to investigate the characteristics causing variations within the degradation behavior by analyzing the similarities of system design within a cluster and the dissimilarities between machines of other clusters. In addition, new machines could be assigned to the best-matching cluster based on their system design specification. Moreover, the information can be used as a covariate during the development of the prognostic algorithm.

4.4.6 Process Extension to Multivariate Time-Series

The presented fleet prognostic type characterization process focuses on the evaluation of a univariate time-series, which is suitable for applications with one dominant physical sensor. However, this is not the case in many applications, especially in view of the increasing complexity of machines and a large number of available sensors (Wang et al. 2012, p. 623). In this case, the fleet prognostic type characterization process must be adapted to the multi-dimensional situation. In order to handle the multi-dimensionality, three approaches are recommended, each of which is associated with strengths and shortcomings.

Individual Sensor Assessment: The simplest approach to handle multi-dimensionality for fleet prognostic type identification is to target relevant sensors individually. This analysis provides a good understanding of each sensor by identifying the degradation behavior of the fleet with regard to one sensor. Thus, differences between the individual sensors are discovered, which could also lead to an improved sensor selection for prognostics. However, the results only indicate the sensor fleet prognostic type, meaning that the individual results of the sensors must then be combined thereafter to determine the fleet prognostic type of the total data set. This step is not trivial in the case of different sensor fleet prognostic types within one data set and requires domain expertise.

Multivariate Distance Measures: Another approach targets the adaptation of the distance measure (step 3 – similarity matching) to handle multi-dimensionality. In comparison to the first approach, data is reduced to a one-dimensional representation during distance calculation, which leads to a simple evaluation of the respective fleet prognostic type. The main disadvantage of this approach is, however, that the behavior of a single sensor might have little influence on the defined fleet prognostic type. The identified fleet prognostic type is strongly characterized by predominant sensor behaviors. This aspect might pose a problem, as a significant deviation within one sensor will have an impact on the accuracy of the prognosis and thus distort the result. In a multivariate time-series distance analysis, the informative value with regard to single sensors is thus limited. Even though sensors might exhibit clear clustering structures, these differences could be weakened and therefore lost during cluster analysis.

Dimensionality Reduction/ Health Indicator Construction: Similar to multivariate distance measures, dimensionality reduction and health indicator construction reduces the data to one or a few dimensions. The most common technique for dimensionality reduction depicts the principal component analysis (PCA), which reduces data through linearly uncorrelated principal components. Health indicator construction, on the other hand, maps data to an artificial health indicator defined in the interval [0, 1]. This is frequently done using linear or logistics regression. The advantage of this approach is the simple fleet prognostic type definition with only one dimension. However, the lack of knowledge on the trajectory behavior within single sensors and their influence on the prediction accuracy are shortcomings, as already mentioned for multivariate distance measures.

Although all approaches are suitable, taking into account their shortcomings, the individual sensor assessment is proposed, given that the information gain on individual sensors is the greatest. However, in this case, the fleet prognostic type assignment may be ambiguous due to conflicting fleet prognostic types of individual sensors. Here, the decision on the overall type should be based on additional application or expert knowledge and might include, among others, the sensor's importance, the availability of data, and the targeted accuracy. Further strategies include a majority vote or assigning the highest fleet prognostic type. As a rule of thumb, it is preferable to assign the highest sensor fleet prognostic type to the complete data set, as methods are applicable to lower fleet prognostic types but are less suitable for higher fleet prognostic types as the respective characteristics are not covered explicitly.

Nevertheless, multivariate distance measures and dimensionality reduction are both valid approaches for a simple and fast fleet prognostic type identification and can also be carried out with regard to the proposed process.

4.5 Evaluation of the Characterization Method

In order to evaluate the proposed characterization method for fleet prognostic, an experimental evaluation is conducted. This is done by means of well-known data sets of turbofan engines (Ramasso and Saxena 2014, p. 1). For this purpose, four data sets, each comprising a fleet of turbofan engines, which were generated under different conditions and settings, are considered. The four data sets are well suited to display different fleet characteristics and types.

The evaluation chapter is structured as follows: in Chapter 4.5.1, the data sets are introduced and classified in accordance with the developed fleet description. Lastly, the fleet identification process is performed to identify the corresponding fleet prognostic types of the four data sets (Chapter 4.5.2).

4.5.1 Fleet Description

The four data sets are published by the NASA Prognostics Center of Excellence data repository (Saxena and Goebel 2008) with the purpose of providing benchmark data sets for data-driven prognostic methods (Ramasso and Saxena 2014, p. 1). In the research field of PHM, the data sets are well-acknowledged and popular and thus have been investigated frequently with varying foci (Elattar et al. 2016, p. 136; Ramasso and Saxena 2014, p. 1). In the following, the four investigated data sets are labeled FD001, FD002, FD003, and FD004 according to the data set description provided.[22]

To gain an initial understanding of the four data sets' characteristics, the underlying fleets are characterized, taking into account the general fleet description as introduced in Chapter 4.4.1. Since the four fleets differ only by a few characteristics, they are jointly introduced, and differences emphasized where present. In the following, the five dimensions and the associated categories for which information was available are discussed successively.

Application Area Dimension

Unit of Analysis: The fleet comprises turbofan engines of commercial aircraft.

Level of Implementation: The data sets target the degradation of the high-pressure compressor (HPC) component of a turbofan engine (Saxena and Goebel 2008, p. 7). The HPC depicts one of the five rotating components within an engine. In addition, there are four additional rotating components, namely the low-pressure compressor (LPC), high-pressure turbine (HPT), and low-pressure turbine (LPT) (Wang 2010, p. 79). The location of the HPC component within the engine can be seen in the schematic view in Figure 35.

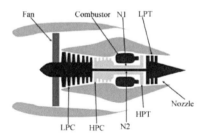

(Frederick et al. 2007, p. 3; Saxena et al. 2008b, p. 2)

Figure 35 | Simplified Engine Diagram of the Turbofan

Fleet Composition Dimension

Size: The number of engines available depends on the respective data set. While FD001 and FD003 comprise 200 machines each, FD002 and FD004 are even larger with 519, respectively, 497 machines. All four fleets can therefore be classified as rather large.

---

[22] The data sets correspond to the sixth data set as provided by the NASA Prognostics Center for Excellence.

System Design: Each engine in the fleet exhibits a different degree of initial wear and manufacturing variations (Saxena et al. 2008b, p. 5). The engine's system designs are therefore subject to variations.

Units/ Functioning: As the engines are installed in different aircraft, they are independent of each other and have an identical function.

Fault Modes: Depending on the data set, the engine can experience one or multiple simultaneous fault modes. FD001 and FD002 exhibit only one failure mode, whereas FD003 and FD004 are subject to two failure modes.

Operating Condition Dimension

Load/ Stress: Depending on the data sets, engines are operated under identical or varying operating conditions. Here, FD001 and FD003 are experiencing only one operating condition, while FD002 and FD004 are subject to six different conditions, which are changing over time. The current operational condition is indicated by three operational settings, presenting the altitude, Mach number (speed), and Throttle Resolver Angle (TRA) value. It is known that the deterioration of the engines is affected by these operating conditions (Ramasso and Saxena 2014, p. 2).

Data Dimension

Structure/ Values: Data is structured and contains continuous-discrete values.

Dimension: Data is multivariate and contains information provided by 21 sensors (Ramasso 2014, p. 5). Measurements include, among others, temperature, pressure, and speed (Wang 2010, pp. 79 f.); however, the meaning of each sensor is not revealed (Rigamonti et al. 2015, p. 4).

Stationary/ Type: Data is non-stationary and consists of raw signals.

Generation: Data is generated using the Commercial Modular Aero-Propulsion System Simulation (C-MAPPS) presented by the NASA Army Research Laboratory (Frederick et al. 2007). The simulation models mimic a complex non-linear system with a multi-dimensional response that closely describes a real-world system (Ramasso and Saxena 2014, p. 2). Faults are randomly initiated during the simulation (Coble and Hines Wesley 2011, p. 7), and deterioration is affected by its usage leading to different engine lifetimes (Saxena et al. 2008b, p. 7; Wang 2010, p. 80). Furthermore, signals are disturbed by varying noise (including process and random measurement noise) to handle the complex multistage noise characteristics observed in real data (Saxena et al. 2008b, p. 5; Rigamonti et al. 2015, p. 4).

Run-to-Failure: The training data sets comprise complete run-to-failure behavior, from the healthy condition through fault initiation over multiple flights to the faulty condition (Wang

2010, p. 80), while for the test data sets, the trajectory is cut off at a random point prior to failure. The RUL ground truth, however, is available. For data sets with different fault modes, no labels are provided (Ramasso 2014, p. 5).

Acquisition Time: Each measurement (row) depicts a snapshot recorded once per flight denoted as a single operational cycle (Saxena et al. 2008b, p. 5).

In general, data set FD001 can be regarded as the simplest and is being used most frequently (Ramasso and Saxena 2014, p. 5). As described, it comprises one operating condition and one fault mode. Contrary to that, data set FD004 comprises six conditions and two fault modes and thus is the most complex. Table 14 provides an overview of the key differential aspects of the four fleets. The conceptual fleet description indicates that the machines within the fleets can have different loads and several failure modes, depending on the data set. In addition, the system design is not identical, but rather variations between machines exist. These initial results imply that the four fleets can be typified as homogeneous or heterogeneous rather than identical fleets. Thus, subsequently, it needs to be identified which fleet prognostic type can be assigned to which data set.

Table 14 | Key Differential Aspects of the Four Fleets

Data Set	Train Trajectories	Test Trajectories	Conditions/ Fault Modes
FD001	100	100	1 / 1
FD002	260	259	6 / 1
FD003	100	100	1 / 2
FD004	249	248	6 / 2

The C-MAPPS data set is suitable for the evaluation of the fleet characterization process due to the simulated engine fleets with common fleet characteristics in terms of machine-to-machine heterogeneity (random variations between machines) as well as other influencing factors on the machine's degradation behavior (operating conditions and multiple fault modes). With regard to their different characteristics, the identified fleet prognostic types are most likely represented in the C-MAPPS data set. Furthermore, the data sets are designed for data-driven prognostics approaches due to little system information available, a big data volume and thus enables large generalizability of the developed procedures (Ramasso and Saxena 2014, p. 2). For fleet characterization, only the train trajectories are investigated as they exhibit complete run-to-failure data to enable fleet prognostic type definitions.

4.5.2 Fleet Prognostic Type Identification

The fleet identification process is carried out in the following for the four fleets introduced above. For this purpose, each step is outlined, and the detailed design, as well as the results achieved with regard to the four fleets, are displayed. As there were often several design options available, as outlined in Chapter 4.4.2, multiple configurations were implemented and explored.

To conclude, an analysis of the impact of the design options was performed. Figure 36 summarizes the performed steps and the specific implementation details, which have been considered within the experimental evaluation.[23]

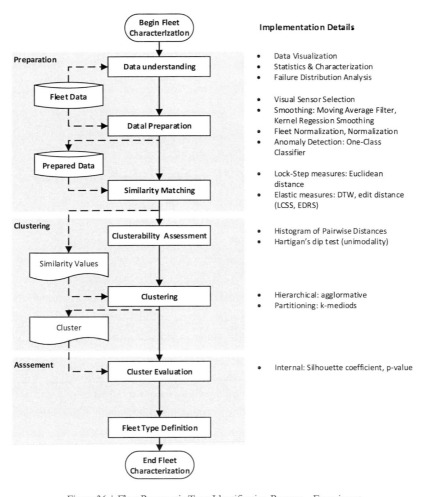

Figure 36 | Fleet Prognostic Type Identification Process – Experiment

Data Understanding | In order to get familiar with the four data sets, data is visually inspected, and simple statistics are performed. Table 15 describes the main results in terms of the minimum and the maximum number of cycles (i.e., number of flights) in the fleet for each data set, as well as a general visual signal trend analysis of the degradation behavior. It can be seen that the sensors among the different data sets in general exhibit a similar trend behavior. While for the

[23] The specific steps are conducted and implemented in Python 3.7 using several additional python libraries. In addition, R statistics was used for an impact analysis due to available packages for full factorial experimental design. A list of all used packages (for both experiments) is within provided in Table 40 in Appendix I.

125

first two data sets, most sensor signals exhibit in- or decreasing trends, for data sets 3 and 4, more sensors have ambiguous trends.

Table 15 | Data Understanding – Simple Analysis

Data Set	Number Cycles		Sensor Trend Analysis		
	Min	Max	No Trend	In-/ Decreasing Trend	Varying Trend
FD001	128	362	1, 5, 6, 10, 16, 18, 19	+: 2, 3, 4, 8, 11, 13, 15, 17[3] -: 7, 12, 20, 21	9, 14
FD002	128	378	1, 5, 6, 10, 16, 18, 19	+: 2, 3, 4, 11, 15, 17[3] -: 7, 12, 20, 21	8, 9, 13, 14
FD003	145	525	1, 5, 6 [1], 10, 16, 18, 19	2, 3, 4, 8[2], 11, 13[2], 17[3]	7, 9, 12, 14, 15, 20, 21
FD004	128	543	1, 5, 6, 10, 16, 18, 19	2, 3, 4, 11, 17[3]	7, 8, 9, 12, 13, 14, 15, 20, 21

Legend: (1) Completely different behavior per machine, however no general trend, (2) Initially distorted signal, (3) Discrete increasing values

With regard to data visualization, the trajectories of each sensor and machine in the fleet are displayed superimposed with the cycle number. To do so, the cycle number was adjusted and negated based on Wang et al. (2008, p. 4) to better fit as remaining useful life label, such that $t = 0$ defines the last cycle per machine before failure and all previous cycles are denoted by $t < 0$. Figure 37 exemplarily shows the signal behavior of data set FD003 considering the adjusted cycle label. After plotting and analyzing all data, five different sensor behaviors are observed in the data (representative sensors are shown in Figure 37). Sensor 1 is illustrative of the constant (no-trend) behavior. This sensor type takes one or very few discrete values for all machines in the fleet. Nevertheless, no trend can be observed. Sensor 11 represents the increasing trend behavior. It can be seen that all machines exhibit a clear trend when approaching $t = 0$. For sensor 14 on the other hand, variations in signal towards the end can be recognized, yet there is no clear increasing or decreasing trend among all machines in a fleet. The fourth behavior, depicted by sensor 17, examplifies the discrete increasing trend. Lastly, both increasing and decreasing trend is displayed in the plot of sensor 20.

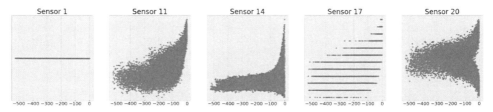

Figure 37 | Data Visualization – Sensor Analysis (FD003)

Another aspect that is easily noticeable during visual analysis is the effects of the varying operating conditions in data set FD002 and FD004. Depending on the current operating condition, sensor values differ considerably (cf. Figure 38). This effect is seen for all sensors in the two data sets. Thus, in order to handle these data sets, the effect needs to be handled during data preparation.

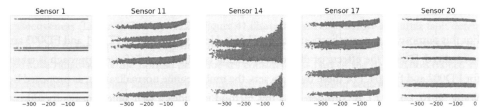

Figure 38 | Data Visualization – Effects of Different Operating Conditions (FD002)

In addition to data visualization, the failure time distribution has been analyzed for multi-modality. Figure 39 portrays the four distributions regarding the four data sets. It is apparent that failure times are distributed within a certain interval, with no indication of multimodality. Thus, from these distributions, no evidence is given for fleet heterogeneity with regard to the engine running time. Nevertheless, the sensor behavior should be analyzed subsequently.

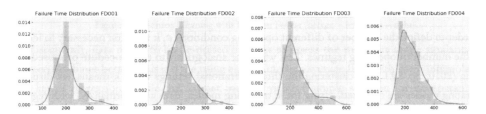

Figure 39 | Failure Time Distribution

Data Preparation | The data understanding step revealed several characteristics of the data sets that must be considered in the data preparation. These characteristics are the initial signal distortion of some sensors as well as the effects of different operating conditions. For that purpose, the three generic tasks of data preparation, namely data and sensor selection, data cleaning, and transformation, as well as fault detection, are taken from Chapter 4.4.3 and detailed for the current case.

Therefore, the first generic task, the data and sensor selection, is addressed initially. For this purpose, relevant sensors are selected based on empirical observations during data visualization. Due to the lack of knowledge of the different sensors, all sensors that exhibit a trend are selected for fleet characterization. This includes all sensors in the columns 'in-/ decreasing trend' and 'varying trend' of Table 15 and excludes sensors in column 'no trend'. This approach is less restrictive compared to many other prognostic approaches using the same data set since only sensors are chosen with clear increasing or decreasing trends (e.g., Hu et al. 2012, p. 127; Ramasso 2014, p. 9).

(cf. Chapter 4.4.3). Parameters for the measures LCSS and EDR have been defined in accordance with the general results by Vlachos et al. (2002, p. 8) and Chen et al. (2005, p. 3). In addition, time-series are left-truncated with regard to the shortest time-series for Euclidean distance calculation.

Clusterability Assessment | In order to evaluate the clustering tendency, both a simple histogram of distances as well as the dip-test was implemented. Since the dip-test requires a continuous distance distribution, it was only possible to execute the statistical test for the Euclidean and DTW distance measure. The p-value of the dip-test was considered in order to assess the null hypothesis (uniform distribution of distance values). The null hypothesis was rejected with p-value < 0.05.

Clustering | Given the distance measure, clustering is performed. For this purpose, agglomerative hierarchical clustering and k-medoids are implemented, as they present established clustering methods for time-series clustering with unequal length.

Cluster Evaluation | Only internal measures are considered in the evaluation as no ground truth or labels are available. The fleet prognostic type evaluation, therefore, mainly relates to the Silhouette coefficient and the p-value provided by the cluster tendency assessment. Due to different design options, combined measures are considered for evaluation. This includes mean and median Silhouette coefficient per sensor and data set, their standard deviation and mean absolute deviation, as well as mean and median p-values.

Fleet Prognostic Type Identification | By analyzing the obtained p-values and Silhouette coefficients, both homogeneous and heterogeneous sensors are identified. For data set FD003 and FD004, five sensors each exhibit a high Silhouette coefficient ($SC > 0.7$) and small p-value ($p - value < 0.05$), whereas all other sensors resulted in medium Silhouette coefficients ($0.25 < SC \leq 0.7$) and high p-values. This result reflects the visual observation. The ten sensors identified as heterogeneous clearly show two different behaviors with increasing and decreasing trends, whereas this has not been the case for the remaining sensors. Exceptions to this are sensor 8 and sensor 13 of the data set FD003. While they have a medium Silhouette coefficient with $SC \approx 0.4$, the obtained p-value indicates a possible clustering structure. For this purpose, a deeper analysis is performed. Through visualization of the sensors and the obtained clusters, deviant behavior was observed close to the end of the life cycle. As the time prior to failure is particularly important for prognostics, these sensors are also marked as heterogeneous.

Given this information, the following fleet prognostic types can be assigned:

- FD001/ FD002 - Homogeneous Fleet: all sensors indicate homogeneous behavior

- FD003/ FD004 - Heterogeneous Fleet: several sensors indicate heterogeneous behavior with clearly separable trajectory behavior

This assignment is in accordance with the presented characteristics of the data sets. While the varying working conditions have been normalized through multi-regime normalization, multiple fault modes (FD003 and FD004) resulted in distinct trajectory behavior and by thus, a hetero-geneous fleet. By further analysis of the remaining sensors in FD003 and FD004 with respect to the two clusters identified in the fleet identification process, the distinct behavior of each cluster can be recognized within all sensors. Figure 42 presents four types of identified clustering structures in the two heterogeneous data sets. The first type (depicted by sensor 7) comprises sensors identified by the above process with clearly distinctive behavior. The second type (rep-resented by sensor 8) covers the sensors with a significant p-value but medium SC. Here, the general trend is similar, yet, the shape of the trajectories differs. The third type (visualized by sensor 9) reveals a hidden cluster structure. While one cluster of machines exhibits diverse be-havior close to the end of the life cycle, the other cluster features very similar patterns. Finally, for the fourth type (shown by sensor 11), no groups are identifiable, yet, the clusters are each located at the other end of the spectrum. Given the fleet description (cf. Chapter 4.5.1), it can be inferred that the heterogeneous fleet type results from the existence of multiple fault modes. As the information on the fault mode is not available, the machines could be labeled according to their cluster assignment.

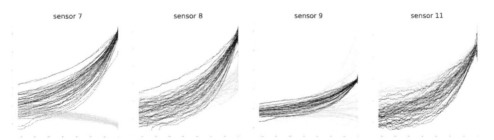

Figure 42 | Fleet Assessment – Identified Cluster Structure (FD003)

In general, with regard to the fleet prognostic type assignment, it was observed that the cluster-ability assessment in the form of the Dip test and the corresponding p-value provides a very good indicator about the quality of the clustering result with respect to possible sub-structures. While for both types 1 and 2 of the identified cluster structures, the null hypothesis was rejected (p-value < 0.05), even type 3 exhibited a mean p-value of approx. 0.4, which is noticeably smaller as for the remaining sensors with p-values around 0.95. A disadvantage, however, is that the dip test is used for continuous data and is hence only applicable for the two distance measures Euclidean distance and DTW. Thus, for the other distance measures, a different test statistic could be considered. Furthermore, the Silhouette coefficient provides support during fleet prognostic type identification.

In order to compare the individual sensor assessment with other approaches (cf. Chapter 4.4.6), fleet identification using the dimensionality reduction method, principal component analysis (PCA), has been carried out subsequently. In contrast to the individual assessment, cluster analysis with PCA did not reveal any structure for neither of the data sets. This leads to the conclusion that individual assessment of sensors should, in general, be favored over dimensionality reduction through sensor aggregation.

Following the identification of the fleet prognostic type, it is of interest to determine whether and to what extent the selection of design options affects the results obtained. For this purpose, the impact of design options on the Silhouette coefficient is analyzed in depth. The experiments are conducted using a *full factorial design*, with three factors at two (smoothing: kernel regression and moving average; clustering method: hierarchical clustering and k-medoids) and four levels (distance measure: Euclidean distance, DTW, LCSS, and EDR). For the analysis, each sensor in the four data sets is investigated separately, leading to 56 experiments with 16 runs, each with different factor combinations.

The analysis is performed using a one-way analysis of variance (ANOVA) and the Kruskal-Wallis Rank Sum Test. Both statistical tests examine whether variability in the response variable is caused by the different levels of a factor. Former depicts a parametric test that requires normally distributed data and homoscedasticity. In order to test both assumptions, the Kolmogorov-Smirnov test (normality assumption) and Levene's test (homoscedasticity) were applied. Results have shown that both assumptions cannot be confirmed for all sensors. Even though various studies indicate that ANOVA is robust against its normal distribution assumption (Blanca et al. 2017, p. 554; Schmider et al. 2010, p. 150; Lix et al. 1996, p. 612) and homoscedasticity with equal group sizes (Glass et al. 1972, p. 273), the comparable non-parametric Kruskal-Wallis test (Glass et al. 1972, p. 237) is additionally executed. Because non-parametric methods come with the disadvantage of information loss due to rank transformation (Blanca et al. 2017, p. 555), both tests are performed for each factor-sensor combination to get a more robust result.

The resulting p-values draw a clear picture. While the investigated smoothing and clustering method do not have any significant impact on the variance of the Silhouette coefficient, the distance measures frequently result in statistical differences among mean values of the groups (levels). Table 16 summarizes the results with regard to the number of tests with significant deviations (p-value ≤ 0.05). It can be seen that for almost three fourth of the sensors, the distance measure presents a significant impact on the resulting Silhouette coefficient with regard to both test statistics, while for smoothing and clustering, this only seems to be the case for few experiments.

Table 16 | Summary Impact Analysis

Statistical Test	Factor Analysis (Number of Tests with p-value ≤ 0.05)		
	Smoothing	Distance Measure	Clustering Method
ANOVA	13	47	3
Kruskal-Wallis Test	1	45	0

In addition, the individual sensor impact is considered. Figure 43 presents the main effect plot as well as the box plot for the distance measure of sensor 1 in data set FD001. A number of observations can be recognized, which are generalizable by visual analysis of the remaining sensors. The main effects plot (left plot) depicts the arithmetic mean for any of the defined levels considering the Silhouette coefficient. For the smoothing option and the clustering method, only minor differences in the mean value of the Silhouette coefficient are observable. Nevertheless, for a large part of the sensors, Kernel regression (KR) and hierarchical clustering (H) results in slightly higher Silhouette coefficients as compared to moving average smoothing (MA) or k-medoids clustering (K). Variations in the mean value, however, do not strongly impact the result. On the other hand, there are significant divergences with regard to the distance measure. The right plot, therefore, details the four levels of the factor distance measure using box plots. Here, it can be concluded that, in particular, LCSS and EDR are responsible for the high variance, while EUC and DTW exhibit little variability. This leads to the conclusion that EUC and DTW are more robust to other design options. The weaker performance of LCSS and EDR could, however, be attributed to default parameter settings. Another observation suggests that DTW commonly causes higher Silhouette coefficients as compared to EUC (with the exception of one sensor), but no consistent pattern is apparent in the other distance measures.

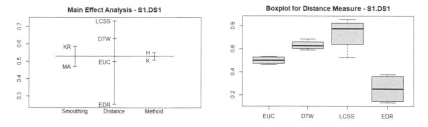

Figure 43 | Detailed Impact Analysis for Sensor 1 in DS FD001

The factor impact analysis indicates a strong influence of the distance measure on the resulting Silhouette coefficient. For this reason, the distance measure should be carefully selected. This finding complies with further experiments carried out in time-series clustering (e.g., Wang et al. 2013, p. 305). It is therefore suggested to use multiple distance measures to increase the overall robustness. Individual Silhouette coefficients, as well as their average, will provide more insights into the fleet structure. Furthermore, well-established measures, such as Euclidean distance and DTW, should be considered as they result in low variance and are thus more robust to design options.

In summary, the evaluation demonstrated that the fleet characterization method yields supportive and structured information about the fleet and the available data. The obtained fleet information can be used subsequently to design an appropriate prognostic algorithm. In a first step, the general fleet description provided a good means to structure and present available knowledge of the four fleets considered in the analysis. Following this, the fleet prognostic type identification process revealed valuable insights in the four fleets and enabled the assignment of fleet prognostic types, which are aligned with visual assessment as well as through described measures and characteristics. In particular, hidden structures were detected for the two fleets (FD003 and FD004) that have not been apparent without the analysis. The conducted evaluation presents an instantiation of the developed method. Its validity was, therefore, only exemplified on the four analyzed fleets. However, the results have been very promising. In order to draw an improved conclusion on the general validity of the method, further instantiations taking into account different fleets, respectively different characteristics are required.

4.6   Discussion and Limitations

The integration of fleet data for prognostics enables an extended database but also faces many challenges due to its heterogeneous nature. The presented method facilitates the characterization of a fleet, both conceptually and data-driven, in order to identify the specific requirements that must be addressed in fleet prognostics. In addition to descriptive fleet characteristics and a data-driven fleet prognostic type delimitation, a process was developed to analyze the composition of the fleet, taking into account the available data. These three results, therefore, satisfy the derived requirements, as presented in Chapter 4.1.

The developed method contributes to an improved understanding of the fleet concept through conceptual and fleet data analysis. Therefore, an existing fleet prognostic type definition has been modified and extended from a data-driven perspective. The newly proposed fleet prognostic types are tailored to the identified fleet heterogeneities and sources of dissimilarity. Even though the majority of identified sources are specific to fleets, some apply for individual machines as well (e.g., multiple faults, dynamic working conditions). Thus, concepts could be transferred. In general, the fleet features are limited to the ones addressed in the current literature. However, as further data analysis is performed, the specific sources of dissimilarity have a negligible impact on the fleet prognostic type definition. In addition, the developed method is easily adaptable to new fleet insights and fleet prognostics research results. Furthermore, the characterization method is targeted towards prognostics. However, ideas and concepts might be transferable for fault detection and diagnostics, respectively.

Although the respective fleet prognostic type can be specified to a certain degree by data visualization and expert knowledge, this intuitive and possibly biased approach could lead to an inaccurate classification. Therefore, a novel data-driven fleet identification based on time-series clustering was developed to improve the reliability of the assignment and identify possibly unknown or hidden structures. Furthermore, it can be used to verify expert knowledge and to quantify the expert's gut feeling. The presented data-driven process facilitates the assignment of a fleet prognostic type by analyzing the similarity of data. Beyond this, helpful insights into the available data can be gathered, and requirements are derived, which will have to be addressed in prognostics (cf. Chapter 5). Nevertheless, while the process performs well on delimitating the homogeneous fleet from the heterogeneous fleet considering the trajectory behavior, it provides only an indication of the identical fleet and can thus be used only in conjunction with expert knowledge regarding validation. However, to strengthen the assignment, a statistical test for white noise or random walks can be performed by learning a general degradation behavior through, e.g., curve fitting and evaluating the resulting residuals. Another limitation of the developed process is its sensitivity to design options, in particular the distance measure, as well as to the cluster's degradation behavior. Former could be mitigated through investigation of multiple measures and their configurations. On the other hand, the recognition of cluster structures is deemed difficult for certain fleet degradation behavior. An example thereof would be clusters that exhibit similar patterns with high intra-cluster variance. It is, thus, advisable to conduct an accompanying visual analysis as well as expert consultation and collate the findings.

To this end, the fleet characterization method for fleet prognostics enables the conceptual description of a fleet, as well as the data-driven fleet prognostic type identification. From this, general data characteristics and requirements, as well as fleet-specific prognostics requirements, can be derived. In particular, the identical fleet demands a prognostic approach to capture random or systematic noise between trajectories, while the homogeneous fleet and the heterogeneous fleet require the integration of further influencing factors, e.g., via covariates and implicitly. Furthermore, given that the heterogeneous fleet is characterized by a clearly separable degradation behavior between more homogeneous subgroups, these sub-fleets should either be addressed individually or learned from an algorithm within prognostics. Based on these requirements, the selection of an appropriate fleet strategy or prognostic method, as well as the development of the fleet prognostic algorithm, will be addressed within Chapter 5.

5   Algorithm Development Method for Fleet Prognostics

On the basis of the identified fleet characteristics (cf. Chapter 4), a process for the development of a prognostic model must be devised that specifically addresses the identified requirements of fleet data. To realize this, an algorithm development method for fleet prognostics is devised, which provides structured and detailed guidance. The method supplements the generic process steps of the PHM design phase of the process reference model for predictive maintenance (cf. Chapter 3) with regard to the processing of fleet data and by providing detailed guidance of the presented process steps.

Thus, this chapter addresses the development of an algorithm development method for fleet prognostics. To achieve this goal, first, the objectives and requirements are derived (Chapter 5.1). Following this, the methodological approach for the elaboration of the method is presented (Chapter 5.2). The resulting knowledge base in form of state-of-the-art fleet prognostics is described in Chapter 5.3. Building on this, the algorithm development method is introduced (Chapter 5.4). Finally, the results of the experimental evaluation are presented (Chapter 5.5). The chapter concludes with a discussion of the results and their main limitations (Chapter 5.6).

5.1   Objectives and Requirements Definition for the Algorithm Development Method

Fleet data has been exploited in prognostics for a long time. Although fleet data are commonly the basis for the development of a global model, the consideration of machine-specific characteristics and degradation behavior has moved into the focus of research only recently (Zhou et al. 2016, p. 13). This analysis offers the advantage that for a more accurate and tailored remaining useful life estimation both the effects of individual usages and the environment can be integrated instead of relying solely on the average fleet behavior (Tsui et al. 2015, p. 2). Thus, the challenge for fleet prognostics is the translation of fleet characteristics into one or a few models so that the data is fully leveraged, yet characteristics of individual machines are adequately addressed. For this purpose, requirements for fleet prognostics are derived in Chapter 4 with respect to three fleet prognostic types. These requirements should be taken into account when developing a prognostic algorithm. Due to the immaturity of the research field, however, current research is focused on building algorithms for specific fleet applications. Therefore, to make research more applicable to other fleet applications, it is necessary to create an application-independent view of fleet prognostic algorithm development that builds on the derived fleet prognostic types. To this end, a method is required which supports the development of fleet prognostic algorithms from data preprocessing to performance evaluation. The method is intended to design an algorithm considering existing fleet data and application characteristics. To achieve this objective, existing fleet prognostic algorithms have to be analyzed in detail and the underlying principles abstracted. From this, four requirements can be defined for the design of an algorithm development method for fleet prognostics:

Fleet Prognostic Requirements: The algorithm development method should accommodate the requirements derived for the three fleet prognostic types. In this way, the method should enable the systematic handling of fleet characteristics.

Prognostic Method Selection Support: The algorithm development method should facilitate the selection of an appropriate prognostic method. For this purpose, it should propose a multi-criteria decision space that embraces data and application characteristics.

Application-Specific Fleet Prognostic Algorithm: The algorithm development method should enable the development of fleet prognostic algorithm taking into account application-specific characteristics.

Algorithm Development Processes for Fleet Prognostic: The algorithm development method for fleet prognostics should provide guidance through processes, which define a set of activities and their sequence. The processes should be designed to provide suggestions for the design of the activities.

While the first three functional requirements can be concluded from the research field of fleet prognostics, the fourth structural requirement follows from the targeted artifact.

The first functional requirement draws on the results of the characterization method of Chapter 4. The identified fleet prognostic requirements have to be reflected by the method. For this reason, existing fleet approaches are to be examined for their compliance and mapped to the three fleet types. However, besides the fleet characteristics, further application characteristics are to be considered during prognostic method selection. This is essential as there is no general superiority of a method, and the suitability depends largely on the compatibility with the specific application (Goebel et al. 2017, p. 66). For this purpose, the second requirement entails a conceptual assessment of fleet approaches and prognostic methods. Since the method selection is highly complex and depends on a multitude of criteria (Sikorska et al. 2011, pp. 1803 f.), the requirement calls for multi-criteria decision support for prognostic method selection that covers data and application characteristics. The selected fleet approach and prognostic method should further be leveraged by the method to support the development of an application-specific prognostic algorithm (third requirement). While the method is intended to provide an application-independent view of fleet prognostics, its overall goal is to facilitate the development of an application-specific fleet prognostic algorithm. Finally, the fourth requirement demands structured and detailed support by means of various processes, which should be provided to the user. In addition to the definition of activities and their flow, further design guidance should be offered by the method (Winter et al. 2009, pp. 7 f.). The resulting algorithm development method is, hence, designed to facilitate the implementation of an application-specific fleet prognostic algorithm taking into account the fleet prognostic requirements.

5.2   Methodological Approach

The main objective of the algorithm development method is to analyze and structure the usage of fleet data for prognostics. For this purpose, the method is developed based on the *six-step method development process* shown in Figure 44, which modifies the procedure for the fleet characterization method (cf. Chapter 4.2 Figure 27). As the topic has already been conceptualized in the previous chapter, this step is skipped. The development of the method is grounded on material gathered through a systematic literature review. The methodological approach for the design of the algorithm development method initially requires a precise problem definition and requirements formulation (as described in Chapter 5.1). Thereafter, a systematic literature review considering the approach by vom Brocke et al. (2009) is performed. This approach enables a rigorous, exhaustive, and reproducible literature search. Considering the body of identified literature, results are analyzed and synthesized. This is followed by the construction of the fleet prognostic algorithm development method. To assess its validity, results are evaluated with regard to experiments on the turbofan engine degradation data set.

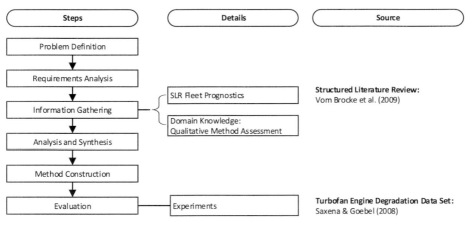

Figure 44 | Methodological Approach (RQ3)

To ensure rigor in the systematic literature review, the search process requires a structural approach and thorough documentation (vom Brocke et al. 2009, p. 6). Primarily, the scope of the review has to be defined, which is accomplished by respecting Cooper's taxonomy (Cooper 1988, p. 109; vom Brocke et al. 2009, p. 7).

Table 17 highlights the defined research scope for this analysis.

Table 17 | Taxonomy for SLR (following Cooper 1988, p. 109)

Characteris-tics	Categories			
Focus	Research Outcomes	Research Methods	Theories	Applications
Goal	Integration		Criticism	Identification of Central Issues
Perspective	Neutral Representation			Espousal of Position
Coverage	Exhaustive	Exhaustive with Selective Citation	Representative	Methodological
Organization	Historical		Conceptual	Methodological
Audience	Specialized Scholars	General Scholars	Practitioners or Policy Makers	General Public

The scientific database Scopus is queried using a combination of the keywords '*prognostics*', '*fleet*', and '*method*', including various synonyms, to retrieve the targeted scientific publications.[24] In contrast to the approach defined by vom Brocke et al. (2009), where the search is concentrated on pre-defined peer-reviewed journals to increase quality and maturity, all journals are analyzed, excluding only non-relevant areas (e.g., medicine, psychology). This is essential given the multi-disciplinary and premature nature of the field. In addition, to expand the knowledge base with relevant conference papers, the PHM society and the IEEE PHM conference processings are examined for fleet prognostic publications. The retrieved papers are analyzed and examined for their suitability. Due to the great amount of identified publications, first, titles are scanned exclusively and irrelevant papers are excluded. This is followed by an abstract analysis. For the most promising papers, a full paper read is performed. Requirements for further consideration are defined as (1) the implementation of a data-driven approach, which (2) utilizes machine condition data (type-III prognostics), and (3) explicitly addresses fleet characteristics instead of learning a global model. The body of literature is further extended through forward and backward search as well as suitable papers identified within the literature review for the characterization method. The search results in a total of 46 papers, which cover 33 distinct approaches.

Besides the systematic literature review, the knowledge base is extended for qualitative method assessment (second requirement). For this purpose, well-known and established prognostic method survey papers are gathered and analyzed regarding described advantages and disadvantages. In addition, further general method descriptions are identified, and relevant information is extracted. The assessment of the methods is subsequently conducted on the basis of criteria from the literature and the information available. As assessment criteria, in particular, criteria defined by Atamuradov et al. (2017, p. 23), Kan et al. (2015, p. 12), Sikorska et al. (2011, pp. 1808 f.), Kotsiantis et al. (2007, p. 263), and Venkatasubramanian et al. (2003, p. 338) are considered. Through an iterative approach, which compares the assessment criteria with the

---

[24] The key words and the final search process is depicted in Figure 74 in Appendix A.

gathered advantages and disadvantages, a subset of suitable criteria are derived, for which information is available. In addition, further application-specific criteria are identified, which include data dimensionality, handling right-censored data as well as system assumptions.

5.3   Information Gathering

Fleet prognostic approaches explicitly address one or multiple fleet characteristics (as identified in Chapter 4.3.2) and thus enable a machine-specific degradation assessment based on the knowledge from the fleet. The identified fleet approaches can be classified into general fleet strategies and methods with fleet capabilities. While the former class depicts strategies that can be configured by different methods, the latter describes methods that handle fleet characteristics inherently or can be adapted for this purpose.

Five *fleet strategies* have been derived from the identified fleet prognostic literature. The (1) local model approach aims to identify similar sub-fleets. For each sub-fleet, a model is built. Finally, prognostics is performed with the model assigned to the respective sub-fleet. The (2) similarity-based learning approach analyzes similarities between machines of fleet and infers an estimate based on this information. Moreover, the (3) correction model approach implements two models. First, the general fleet degradation behavior is targeted, which is adjusted by a second model to capture the machine-specific characteristics. Furthermore, the (4) Bayesian inference and mixed model approach extends any degradation model by means of stochastic parameters, which are calculated individually per machine via Bayesian inference, while deterministic parameters are estimated with respect to the fleet. Lastly, the (5) ensemble learning approach as a meta-strategy combines various estimates of trained models (often different fleet prognostic approaches are considered) and thus, leverages the strength of several methods.

Beyond that, three *methods with fleet capabilities* are identified in the literature. Most commonly applied are (1) artificial neural networks (ANNs). Through a proper selection of features, ANNs can learn different behaviors and are, therefore, able to handle machine-to-machine differences either based on covariates or available sensor data. In addition, (2) Markov models can be adapted to identify various states of degradation resulting from variations within the fleet. The last method depicts (3) Gaussian process regression, which is extended by an additional covariance to capture correlations between the machines.

Table 18 summarizes the identified literature with regard to the assigned fleet approaches as well as method specifications.

Table 18 | Concept Matrix - SLR Fleet Prognostics

Fleet Approach		Source	Method Specification
Fleet Strategies	Local Models	Li et al. 2015	Support Vector Regression
		Peng et al. 2012	Echo State Network
		Yang et al. 2017	Support Vector Regression
	Similarity-based Learning	Bleakie and Djurdjanovic 2013	Gaussian Mixture Models
		Wang et al. 2008; Wang 2010; Lam et al. 2014	Direct via Curve Matching
		Guepie and Lecoeuche 2015	Direct Matching
		Zio and Di Maio 2010; Di Maio and Zio 2013; Eker et al. 2014	Direct via Fuzzy Approach
		Xue et al. 2008	Direct via Truncated Bell Function
		Yu et al. 2012	Direct via Grey Correlation Analysis
		Ramasso 2014	Direct considering imprecise HI
		Wang et al. 2012; Voisin et al. 2013	Direct via Curve Matching
		Chang et al. 2017	Direct via Relevance Vector Machine
		Bektas et al. 2019	Direct
		Bagheri et al. 2015	Fleet Features
		Palau et al. 2020	Parameter Estimation
		Liu et al. 2007	Match Matrix (ARMA)
		Gebraeel et al. 2004; Huang et al. 2007	Feedforward Neural Network
	Correction Model	Liu and Zio 2016	Fuzzy Similarity (Support Vector Machine)
		Trilla et al. 2018	MLP (Linear Regression)
	Bayesian Inference & Mixed Models	Wang et al. 2018	Wiener Process
		Coble and Hines Wesley 2011; Shumaker et al. 2013	Polynomial Model
		Zaidan et al. 2013	Polynomial Model
		Gebraeel et al. 2005; Gebraeel 2006	Exponential Model
		Zaidan 2014; Zaidan et al. 2015a; Zaidan et al. 2016; Zaidan et al. 2015b	Bayesian Hierarchical Model (Polynomial Model)
	Ensemble Learning	Peel 2008; Lim et al. 2014	Various Artificial Neural Networks
		Al-Dahidi et al. 2017	Hidden Markov Model, Fuzzy Similarity
		Hu et al. 2012	Various Fleet Prognostic Approaches
Methods with Fleet Capabilities	Artificial Neural Networks	Heimes 2008	Recurrent Neural Network with Operating Condition Features
		Wu et al. 2017b	Vanilla LSTM with Dynamic Differential Features
		Cristaldi et al. 2016	Feedforward Neural Network with Confidence Interval
		Babu et al. 2016	Convolutional Neural Network with Operating Condition Features
	Markov Models	Al-Dahidi et al. 2016	Hidden Markov Model (state clustering)
	Gaussian Process Regression	Duong et al. 2018; Duong and Raghavan 2018	Gaussian Process Regression with additional Covariance Matrix

Beyond the approaches identified, there are two additional ways to process fleet data for prognostics. On the one hand, this includes the creation of a global model without consideration of fleet-specific characteristics. This represents a simple variant that might be applicable for the identical fleet. On the other hand, data cleaning can be performed during data preprocessing to eliminate the influencing factors. In both cases, a suitable general prognostic method can be applied instead of relying on fleet approaches. These two alternatives, however, are not further discussed in detail in this thesis as the fleet characteristics are not explicitly integrated within the fleet prognostic algorithm development method.

5.4 Algorithm Development Method Design and Construction

The developed algorithm development method for fleet prognostics is organized into three processes. The first process depicts the fleet characterization process, as presented in Chapter 4. Taking into account the identified fleet prognostic type, a suitable fleet approach and method is selected in the second process, the fleet approach selection process, with the help of a qualitative assessment. The last process, the algorithm implementation process, presents the detailed steps to develop a fleet prognostic model. Figure 45 shows the resulting main process, which is the foundation for the algorithm development method for fleet prognostics.

In the following, the fleet approach selection process is presented and detailed (Chapter 5.4.1). Afterwards, the algorithm development implementation process is outlined (Chapter 5.4.2).

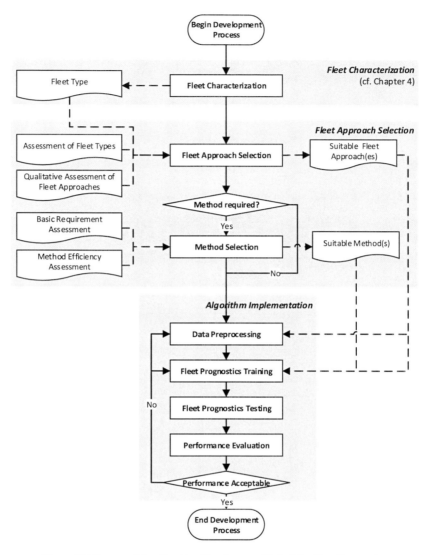

Figure 45 | Process of the Algorithm Development Method for Fleet Prognostics

5.4.1 Fleet Approach Selection Process

The selection of one or more suitable fleet approaches constitutes the first step of the algorithm development method for fleet prognostics. For this purpose, the identified fleet approaches are examined more thoroughly and compared on the basis of different characteristics. The user should, therefore, be provided with a decisive basis for making a suitable selection. The selection of a fleet approach is largely application-specific and can only be performed by evaluating a variety of different criteria. The weighting and prioritization of the different criteria require the involvement of the user based on the application characteristics.

As described above, the selection of a fleet approach covers the determination of a suitable fleet strategy or a method with fleet capabilities. However, since some fleet strategies rely on general prognostic methods, these must also be reflected in the final selection. Thus, the different fleet approaches are first examined and evaluated with regard to their compatibility with the three fleet prognostic types (Chapter 5.4.1.1). Subsequently, a general assessment of established prognostic methods is accomplished (Chapter 5.4.1.2).

5.4.1.1 Fleet Approach Assessment

The five fleet strategies and three methods with fleet capabilities are qualitatively examined in the following for their suitability for different applications, especially considering the three fleet prognostic types. When evaluating the applicability to the fleet prognostic types, the defined fleet prognostic requirements are examined. Approaches suitable for the identical fleet must be able to handle machine-to-machine variations in the form of random or systematic noise between machines. They do not need to incorporate further influencing factors, e.g., through covariates. Approaches for homogeneous and heterogeneous fleets must additionally treat known or unknown influencing factors. For heterogeneous fleets, there are clearly separated subgroups with similar behavior, whereas this is not the case for homogeneous fleets. The allocation of the fleet approaches to the three fleet prognostic types is provided in Table 19. In addition, for each fleet strategy, their ability to handle covariates and the requirement for a general prognostic method are presented. The different strategies and methods are briefly discussed in the following.

Table 19 | Assessment of Fleet Approaches considering Fleet Prognostic Types

Fleet Approach		Fleet Prognostic Type			Covariates Handling	Prognostic Method Required
		Identical	Homoge-neous	Heteroge-neous		
Fleet Strategies	Local Models	-	-	X	Yes	Yes
	Similarity-Based	(X)	X	(X)	No	No
	Correction Model	(X)	X	-	Yes	Yes
	Bayesian Inference & Mixed Models	X	(X)	-	No	No
	Ensemble Learning	(X)	X	(X)	Yes	Yes
Methods	Artificial Neural Networks	-	X	X	Yes	No
	Markov Models	-	(X)	X	Yes	No
	Gaussian Process Regression	X	X	(X)	No	No

Local Model Approach | The local model approach divides the fleet into similar sub-fleets and constructs a separate model for each sub-fleet. For the construction of these local models, every prognostic method can be used. This also includes the fleet approaches for identical and homogeneous fleets. The local model strategy is applicable for heterogeneous fleets, either with or without explicit influencing covariates. It is characterized by its simplicity; however, it requires a sufficient amount of machines within a sub-fleet to build reliable prognostic models. The specific characteristics of this approach highly depend on the selected method(s) for model development.

Fleet Prognostic Type: Heterogeneous fleet with and without explicit influencing covariates

Similarity-based Approach | The similarity-based approach determines the similarity between two trajectories in order to specify the weighting of a trajectory for prediction. The more similar the pattern is, the greater is the weight of the pattern when determining the remaining useful life. The model-based similarity approach denotes a variant, which uses the trajectories similarity to define the parameters of the machine-specific degradation model. This strategy can be applied for all fleet prognostic types; however, it is most suitable for homogeneous fleets. In the case of a heterogeneous fleet, learning should only be limited to the most similar machines. Similarity-based learning requires a high sample size. As the similarity is determined at runtime, this strategy represents a lazy learner. Furthermore, the prediction is performed on univariate data without including covariates.

Fleet Prognostic Type: Homogeneous fleet without explicit influencing covariates

Correction Model Approach | The correction model approach is based on two models, the general degradation model and the correction model. By means of the degradation model, a general prediction is generated initially. This global estimate is adjusted by the deviation learned in the correction model. Since this approach determines deviations due to general behavior, it is best suited for the homogeneous fleet but can also be carried out for identical fleets. The characteristics are strongly dependent on the selected methods. This also applies to the incorporation of covariates.

Fleet Prognostic Type: Homogeneous fleet with and without explicit influencing covariates

Bayesian Inference & Mixed Model Approach | In the Bayesian inference & mixed model approach, a general degradation behavior is given by expert knowledge, and the parameters are adapted to the machine-specific real-time data using Bayes' theorem. This approach does not adequately address other influencing factors and is, therefore, best applied to the identical fleet. In addition, if the functional degradation behavior of the homogeneous fleet is the same, the variant of Bayesian hierarchical models can be used for this type, which considers different levels of influence. Bayesian inference and mixed models can be built on a small sample size; however, they require high computational effort as well as expert knowledge.

Fleet Prognostic Types: Identical fleet without explicit influencing covariates

Ensemble Learning Approach | Ensemble learning refers to the concept of learning multiple models and combining the resulting predictions to improve both accuracy and robustness. The fleet characteristics can be reflected both in the individual models as well as during the fusion of the predictions. Since this strategy heavily depends on the chosen methods and the fusion mechanism, this approach is applicable to all fleet prognostic types. Furthermore, the specific characteristics have to be concretized by the chosen methods. However, sufficient data should be available to construct multiple models.

Fleet Prognostic Types: Identical, homogeneous, and heterogeneous fleet with and without explicit influencing covariates

Artificial Neural Networks | Artificial neural networks can approximate complex patterns due to their multi-layer network structure. This allows them to learn the influence of covariates on the degradation behavior. For this purpose, ANNs can handle both homogeneous and heterogeneous fleets, for which the influential covariates are available. An ANN is a black box method that requires a large amount of data to learn the underlying structures. Moreover, they are computationally demanding, and their parameterization is challenging.

Fleet Prognostic Type: Homogeneous and heterogeneous fleet with explicit influencing covariates

Markov Models | Markov models are dynamic bayesian networks, which model the degradation through a finite set of states. The ability to represent different degradation paths within a model makes them suitable for prognostics with fleet data. Markov models are, therefore, able to specifically capture the characteristics of the heterogeneous fleet both with or without explicit covariates. If the network is trained purely on data, a large amount of data is required. However, right-censored data can also be included in the training. Furthermore, Markov models are very computationally intensive. Their main disadvantage, though, is the underlying Markov assumption.

Fleet Prognostic Type: Heterogeneous fleet with and without explicit influencing covariates

Gaussian Process Regression | The method Gaussian process regression approximates the degradation behavior by a multivariate Gaussian distribution. By means of an additional covariance, the correlation between the different machines can be integrated into the prognosis. Thus, this variant of Gaussian Process Regression is able to capture the degradation behavior of both identical and homogeneous fleets. Gaussian process regressions can be trained on small multivariate sample data without explicit covariates. However, they are time-consuming during learning, require expert knowledge to select a suitable covariance, and are therefore challenging to create.

Fleet Prognostic Type: Identical and homogeneous fleet without explicit influencing covariates

Apart from the evaluation of fleet approaches with regard to the fleet prognostic types, a qualitative assessment is performed.[25] For this, eight criteria are examined in detail. The investigated criteria are clustered in the four groups, namely data-based, complexity-based, user-based, and system-based. The data-based group comprises the criteria of (1) dimensionality and (2) sample size. The first criterion specifies whether the approach requires a univariate data representation

---

[25] A list of all advantages and disadvantages for the fleet approaches is provided in Table 34 in Appendix H.

or whether it is capable of handling multivariate data. The sample size criterion refers to the amount of data available for algorithm development. This criterion, therefore, corresponds to the fleet size, as defined in Chapter 4.4.1. For the complexity-based group, the criteria (3) learning time complexity and (4) run time complexity are analyzed. The former indicates the time duration to train the model (training step), while the latter refers to the time complexity to derive RUL estimates (testing/ online step). The user requirements are emphasized by the user-based group with regard to (5) transparency and (6) parameter handling criteria. The transparency criterion describes the traceability and comprehensibility of the remaining useful life estimation for the user. Approaches with high transparency often increase trust by the user in the prediction. In addition, the parameter handling criterion depicts the complexity of tune parameters during model training. Approaches with low parameter handling usually only have no or very few parameters, which have to be defined, whereas high parameter handling requires extensive parameter optimization. Lastly, criteria within the system-based group are the (7) output type and (8) expert knowledge. The output type defines whether the approach yields a point estimate or a failure distribution. A failure distribution supports the quantification of the prediction uncertainty, while the point estimate reflects the mean expected failure time. The need for additional expert knowledge is defined within the expert knowledge criterion. Models with no expert knowledge are purely trained through the data.

Table 20 shows the resulting qualitative assessment.[26] Given that the characteristics of several fleet strategies depend on the selected prognostic method, it is not possible to evaluate all criteria. In addition, a rating is added for criteria that are evident but have no foundation in literature.[27]

To select an appropriate fleet approach, the subset of possible approaches for each fleet prognostic type is first identified, considering Table 19. In addition, the availability of covariates should be taken into account and included in the selection. The resulting subset of approaches is subsequently evaluated with regard to Table 20. Here, the criterion of sample size should be examined a priori for its suitability. Thereafter, the remaining criteria can be weighted based on their relevance and importance to the application case. The best matching fleet approach(es) are to be considered for the specific application. In the case that the approach requires the definition of a prognostic method, as depicts in Table 19, it is necessary to proceed with the method selection process (Chapter 5.4.1.2). Otherwise, this process can be skipped and directly continued with the algorithm implementation process (Chapter 5.4.2).

---

[26] The basis of assessment for the qualitative assessment of fleet approaches is specified by the numbers in the superscript. An assignment of the number to the sources can be found in Table 35 in Appendix H.

[27] The assessment for the three methods are taken from the general assessment (Chapter 5.4.1.2).

Table 20 | Qualitative Assessment of Fleet Approaches

	Data-based		Complexity-based		User-based		System-based	
	Dimen-sionality	Sample Size	Learning Time Complexity	Run Time Complexity	Transparency	Parameter Handling	Output Type	Expert Knowledge
Local Model	-	High	-	-	High	-	-	-
Similarity Based	UV	High[2,7,17]	Low[1,3]	High[1,3]	High	Low	P	No[1,2,3,5]
Correction Model	-	Medium - High	(High[5])	-	Medium	Medium	P	-
Bayesian Inference	UV	Low	High[8,10,21]	High[8,10,21]	Low[8,10]	High[8,10]	FD[11]	Yes[5,8,10,16]
Ensemble Learning	-	High	High	High	Low	Medium-High	P	-
Artificial Neural Networks	MV[1,2,3,4,5,15]	High[1,2,3,6]	Medium[19] – High[3,4,18]	Medium[1,18] – High[1]	Low[1,2,5,15]	High[1,3,4,23]	P[3]	No[1,3,5]
Markov Models	MV[3,9]	High[1,2,3,4,9,14]	High[1,3,4,14,18]	High[3,18]	Low[2]	High[1,2,9,23]	FD[1,3,9]	Both[1,3]
Gaussian Process Regression	MV[1,2,12]	Low[1,2,12,20]	High[1,2,10,18,25]	Low[18]	High[1,2,13]	Medium[13,26] – High[1,4]	FD[1]	Yes[1,18]

Legend: UV = univariate, MV = multivariate, P = point estimate, FD = failure distribution

5.4.1.2 General Method Assessment

Having analyzed the fleet prognostic approaches, the second step is to present a general assessment of prognostic methods. This is required as several fleet strategies employ general prognostic methods. For this purpose, selection support is also needed for the general prognostic methods. Due to its generality, it can also be used in general prognostic algorithm selection.

The assessment of methods is performed with regard to thirteen criteria. Criteria are separated into basic method requirements and method efficiency. The first class of criteria, basic method requirements, analyzes the different pre-requisites of the methods. Criteria in this class can be interpreted as hard constraints. Only those methods that meet the specific requirements of the application should be analyzed for their efficiency in a second step. This second class thus comprises the criteria of method efficiency. Criteria within this class can be regarded as soft constraints. Based on these criteria, a final selection is made. This could include one or multiple suitable methods. The general assessment covers the common prognostic methods (as specified in Chapter 2.3), namely the simple trend regression, autoregressive models, stochastic models, Markov models, artificial neural networks, support vector machines, and gaussian process regression. Since there are different variants and further developments for each of the methods, only the basic form is examined. Nevertheless, well-known variants and their key differentiating characteristics are summarized as well.

The *criteria of basic requirements* can be further subdivided into data and system requirements. The group of data requirements consists of the criteria (1) dimensionality, (2) sample size, and (3) right-censored. Therefore, these criteria represent the general characteristics of the data. The first criterion, the dimensionality of the data, defines whether the method is capable of univariate or multivariate analysis. Although it is possible to reduce multivariate data to a one-dimensional space, information is lost. Consequently, if no strong health indicator is present, a multivariate analysis should be favored. The second criterion focuses on the sample size that is considered for method training. Methods that are assigned a small sample size can be trained with few degradation trajectories, while methods with a large sample size need a much higher number of trajectories to learn their pattern. Finally, the third criterion, right-censored, evaluates the ability of the methods to learn from data that is right-censored (failure event has not yet occurred). Partial data can contain valuable information that is not captured by methods that are only trained by complete run-to-failure data. In the case of few run-to-failure data (e.g., due to rare machine failures through pre-determined maintenance), a method should be chosen that can learn from right-censored data. The second group within the basic requirements are criteria related to the system requirements, namely (4) output type, (5) expert knowledge, and (6) general system assumptions. The criterion output type specifies the ability of the method to provide confidence intervals. Methods that only supply point estimates provide no information about the associated uncertainty in the prediction. In particular, for critical systems, the additional information provided by the probabilistic failure distribution can support an improved maintenance decision given the uncertainty of the estimate. The fourth criterion, denoted by expert knowledge, considers the necessity of prior expert knowledge for the creation of the model to specify the degradation behavior. In case little knowledge of machine degradation is available, pure data-driven methods are more applicable. Finally, the criterion of system assumptions specifies the assumptions that the method makes about the degradation behavior. For the applicability of a method, the assumptions for the system should be fulfilled. Table 21 provides the final assessment of the methods considering the basic requirement criteria, as well as the frequently discussed variants, including their key differential characteristics.[28]

Table 21 | Basic Requirements Assessment for Prognostic Methods

| | | Basic Requirements | | | | | | Frequently Discussed Variants |
| | | Data Requirements | | | System Requirements | | | |
		Dimension-ality	Sample Size	Right-Censored	Output Type	Expert Knowledge	System Assump-tions	
Statistical/ Stochastic Models	Simple Trend Regression	UV[2]	Low[6,27]	Partly (extrapolation)	P[3] / FD[3]	No[18]	• Monontonic trend[3,5,14]	Linear regression: monotonic[3] Non-linear regression: non-monotonic[29] Multiple linear regressions: multivariate
	AR Model	UV	Low[2,3,4,6,9,18]	Yes	P[9]	No[3,18]	• linear time-invariant[2] • stationary, non-dynamic[3,4]	ARIMA: non-stationarity VARMA: multi-variate
	Stocha-stic Processes	UV[14]	Low[14,24]	Yes	FD[2,14]	Yes[28]	• Markov Property[2] • Independent additive increments	Wiener Process: non-monotonic[14,24] Gamma Process: monotonic[2,14,24]
	Markov Models	MV[3,9]	High[1,2,3,4,9,14]	Yes[1,3,9]	FD[1, 3,9]	Both[1,3]	• Markov property[1,2,3,7,14,16] • Independence assumption[3,7,9,14,16,17] • Finite-state space[1,2,24] • stationary, non temporal[1,3] • Non-linear[1,2]	HMM: unobserved degradation states, non-stationary[1,3,4] HSMM: unobserved degradation states, inconstant failure rate/ temporal structures, non-stationary[1,3,4]

---

[28] The basis of assessment for both the basic requirements as well as the method efficiency is specified by the numbers in the superscript. An assignment of the number to the sources can be found in Table 37 in Appendix H.

Machine Learning	ANN	MV[1,2,3,4,5,15]	High[1,2,3,6]	No[1,2,3,6]	P[3]	No[1,3,5]	• Complex (unknown), unstable or non-linear[1,2,3,4,5,15]	RNN/ ESN/ LSTM: dynamic network FNN/ CNN: static network
	SVM	MV[1,8]	Low[1,9] – Medium[1,2]	No	P[1,10]	No	• Non-linear, non-stationary[1]	RVM: Probabilistic distribution
	GPR	MV[1,2,12]	Low[1,2,12,20]	Yes	FD[1]	Yes[1,18]	• non-linear, non-stationary [1,2,4,12,26] • Gaussian likelihood[1,4,26] • Data values are normally distributed[1]	

Legend: UV = univariate, MV = multivariate, P = point estimate, FD = failure distribution

The second part of the evaluation examines the *efficiency of the method*. Within this class, the identified criteria are divided into the categories accuracy-based, complexity-based, and user-based. The first group of criteria (accuracy-based) comprises the criteria (1) general accuracy, (2) prediction horizon, and (3) robustness. The general accuracy criterion indicates the average performance of the method for estimating the remaining useful life. However, this strongly depends on the suitability of the method and its parameterization. The second criterion, the prediction horizon, provides details about the time span in which a method can make reliable predictions. Robustness, the third criterion, examines the ability of a method to process noise and outliers. The complexity-based criteria can be distinguished into (4) run time and (5) learning time complexity. Here, the amount of time a computer takes to learn a model (learning time) and execute it (run time) is evaluated. As a third group, the user-based criteria comprise the (6) parameter handling and (7) explainability/ transparency. The criterion parameter handling focuses on the simplicity of developing a model, especially considering the determination of the parameters. Lastly, explainability and transparency describe the traceability of the prediction result. This can be valuable in increasing user confidence in the prediction.[29]

The presented selection criteria must be appropriately weighted and prioritized for each application. This is done analogously to the previous chapter by an iterative reduction of the set of suitable methods. For this purpose, first, a subset of suitable methods should be selected with respect to the hard requirements derived in Table 21. The methods that meet the specific application requirements can then be analyzed in terms of method efficiency (Table 22). Here, the application-relevant criteria have to be defined by experts. The best matching method(s) is selected for algorithm implementation (cf. 5.4.2). In general, however, for the development of a

---

[29] A summary of all collected advantages and disadvantages can be found in Table 36 in Appendix H.

prognostic method, an appropriate subset of methods or a single method thereof should be considered. Results of the prognostic method efficiency assessment are provided in Table 22.

Table 22 | Method Efficiency Assessment for Prognostic Methods

| | | Method Efficiency | | | | | | |
| | | Accuracy-based | | | Complexity-based | | User-based | |
		General Accuracy	Prediction Horizon	Robustness	Run Time Complexity	Learning Time Complexity	Parameter Handling	Explainability/ Transparency
Statistical/ Stochastic Methods	Simple Trend Regression	Medium[2,9,18,21]	Short[18]	Low[3,18]	Low[18]	Low[2,9,18]	Low[3,5,6,22,28]	High[2]
	AR Model	Medium[18]	Short[2,3,4,9,18]	Low[3,4,18]	Low [3,18]	Low[3,18]	Medium[15]	High[2]
	Stochastic Processes	Medium[18]	Short[18]	Low[3,28]	Low[18]	Low[18]	Low[3,9,14,28]	High[9,14,28]
	Markov Models	High[4,18]	Long[18]	Medium[18] – High[3]	High[3,18]	High[1,3,4,14,18]	Low[3,22]	High[1,3,23,24]
Machine Learning	ANN	High[2,4,18]	Long[1,18]	High[1,3,18,20]	Medium[1,18] – High[1]	Medium[19] – High[3,4,18]	High[1,3,4,23]	Low[1,2,5,15]
	SVR	Medium[21] – High[1,4,18]	Long[18]	Medium[2,18] – High[1,7,8,9]	Low[1,4,7,8]	Medium[4,7,8,11] - High[8,18,19]	High[1,2,9,23]	Low[2]
	GPR	High[18]	Medium[18]	Medium[1] – High[1,4,18]	Low[18]	High[1,2,10,18,25]	Medium[13,26] – High[1,4]	High[1,2,13]

To conclude, a key finding can be derived from the qualitative assessment of general prognostic methods. By examining the results with regard to similarities and differences in the assessment of the methods, two clusters are differentiated, which exhibit high intra-cluster similarities and inter-cluster differences. On the one hand, the first cluster, which comprises the methods of simple trend regression, autoregressive methods, and stochastic processes, can be attributed to the characteristics of univariate analysis and restrictive system assumptions. On the other hand, the second cluster, including the methods ANN, SVR, and GPR, is less restrictive on the system assumptions, can incorporate multiple variables, and generally requires more data for training. Methods in the second cluster are superior with regard to their general accuracy and robustness. Furthermore, they are more reliable with a longer prediction horizon. However, their time complexity is higher, and they are generally more time-consuming during creation. Markov models represent the intersection between these clusters. While most characteristics are compatible with the second cluster, the system requirements are limiting their usage. Nevertheless, with regard to the different variants, some of the restrictions can be resolved, making them more generally applicable. Regarding the suitability of the methods within the two clusters, it can be said that for a simple and clearly defined degradation behavior, methods of the first cluster are better qualified, whereas, for complex behavior, methods of the second cluster should preferably be applied. In addition, it can also be advantageous to first consider methods from the first cluster to gain a better understanding. This knowledge can subsequently be used to improve the predicted estimate using methods from the second cluster.

5.4.2 Algorithm Implementation Process

The second step of the algorithm development method comprises the implementation of a fleet prognostic algorithm. For this purpose, the fleet approaches are examined in detail, and the relevant steps and configuration options are presented. In contrast to the selection of methods, only the fleet strategies and the methods with fleet capabilities are targeted in detail. The algorithm development implementation can be structured within three processes, namely, data preprocessing, method implementation (including training and testing), and performance evaluation, which are executed iteratively. Figure 46 presents the derived reference view of the algorithm implementation process. The reference view outlines the three processes, along with the key elements for each process. In particular, the derived general steps for each fleet approach are highlighted. The steps marked with dark grey boxes represent the steps required for method training, whereas those marked in light grey reflect both test and online (prediction) steps. Furthermore, the upper fleet approaches visualize the five fleet strategies (uniform background), which is followed by the methods with fleet capabilities (striped background).

The three processes are detailed in the following sub-chapters. First, the data preprocessing steps are presented (Chapter 5.4.2.1). Thereafter, the fleet approaches, as well as their key steps and respective configuration options, are discussed (Chapter 5.4.2.2). Finally, in order to assess the performance of the developed prognostic methods, different performance evaluation metrics are presented (Chapter 5.4.2.3).

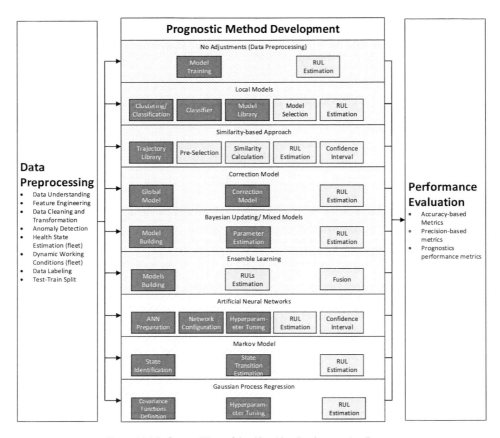

Figure 46 | Reference View of the Algorithm Implementation Process

5.4.2.1 Data Preprocessing

Prior to performing prognostics, data must be preprocessed adequately. This step is of central importance, even though it is sometimes neglected. The quality of data must be reviewed and improved, if necessary. In addition, the data representation has to be aligned to the respective algorithm requirements. This phase highly depends on the available data and the selected prognosis algorithm. Nevertheless, the main aspects and steps are presented here, which should guide the user during the data preprocessing process.

The data preprocessing process comprises various cycles, especially for feature engineering and data preparation, and can therefore not be performed completely sequentially. This is especially true for some sensor selection methods, which require, for example, prior data cleaning (e.g., the Empirical Sensor-Noise Ratio (Wang 2010)) and a health indicator (e.g., the Grey Correlation Analysis (Yu et al. 2012)). The data preprocessing, hence, requires an iterative reduction of the data and its preparation.

Figure 47 visualizes the process for data preprocessing. The steps presented are limited to those identified in the fleet prognostic literature. Nevertheless, they should be representative for the most part. The data preprocessing process is detailed in the following.[30]

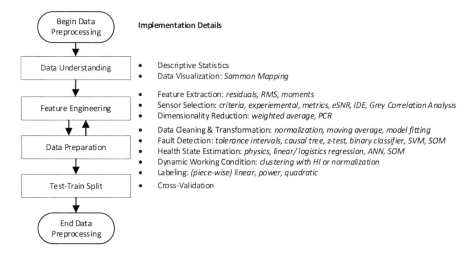

Figure 47 | Data Preprocessing Process

Data Understanding | In order to gain a first impression of the data, the available data needs to be analyzed. This aspect is often not described in detail as authors usually do not consider it as being part of the algorithm presented. Nevertheless, it presents an important element for the following steps. Here, simple *descriptive statistics* for data description (e.g., Heimes 2008) and different *data visualizations* such as Sammon Mapping (e.g., Peel 2008) are examples.

Feature Engineering | Data sets often comprise a huge variety of (sensor) data, not all of which are relevant for prognostics. For this purpose, feature engineering enables the identification of appropriate variables and features. Feature engineering comprises feature extraction, sensor selection, as well as dimensionality reduction. In order to retrieve information from the provided data, additional *features can be extracted*. This step is highly application-dependent (Wang 2010, p. 39) and cannot be generalized. Nevertheless, to name a few options, features can include residuals after fault detection (e.g., Xue et al. 2008), root mean squared values (e.g., Wang et al. 2018), or time-shift invariant moments (e.g., Liu et al. 2007). At the same time, the data set needs to be reduced to a manageable and adequate subset. For this purpose, sensors and variables that show deterioration over time must be identified. The *selection of qualified sensors* is mostly based on data screening. Exclusion criteria are hereby, for example, single or multiple discrete values or non-monotonic trends (e.g., Wang et al. 2008, Wu et al. 2017b). This intuitive selection of sensors is often combined with the execution of multiple experiments with potential pre-defined

[30] An overview of the identified steps and the respective sources is provided in Appendix H in Table 38.

data subsets (e.g., Ramasso 2014). Furthermore, the three metrics, trendability, prognosability, and monotonicity, have been developed to indicate the applicability of sensors for prognostics (Coble 2010). Other systematic approaches for sensor selection include the Empirical Signal-Noise Ratio, which evaluates the local variance of sensors in comparison to its global variance (Wang 2010), the Improved Distance Evaluation, which selects sensors based on their distance between healthy and faulty condition (Bagheri et al. 2015), as well as a sensor selection approach using Grey Correlation Analysis (Yu et al. 2012). Lastly, *dimensionality reduction* can be applied in terms of combining features using weighted average (e.g., Coble and Hines Wesley 2011) or Principal Component Analysis (PCA), which performs a linear mapping to a lower-dimensional representation (e.g., Wang 2010).

Data Preparation | The selected data must be converted into a format required by the different algorithms. This includes cleaning and transforming data, identifying the degradation behavior in terms of fault detection, estimating the health state, handling dynamic working conditions and labeling the data.

High data quality is essential for further analysis. Without initial processing, real-world data is often ascribed to noise and might consist of different scales. For this reason, *data cleaning and transformation* should be conducted prior to the evaluation. To make data comparable and by this remove their scale dependency, data should be normalized (e.g., Wu et al. 2017b). Furthermore, data is generally smoothed to eliminate noise and outliers. For this purpose, moving average depicts a generic and simple procedure (e.g., Riad et al. 2010). Model fitting represents another alternative, though it depends highly on the degradation behavior (Wang 2010, p. 42). For this, options include exponential fitting as well as the usage of support vector machines or relevance vector machines (e.g., Hu et al. 2012).

Prognostics examines the degradation behavior. While some machines degrade continuously, others operate in normal conditions until a fault is initiated. In the second case, prognostics should focus on the degradation part of the time-series. For this purpose, *fault detection* approaches are applicable. In the simplest case, tolerance intervals are defined, and deviations are considered as anomalies (e.g., Zio and Di Maio 2010). In addition, there are manifold other more sophisticated alternatives, including among others the causal tree (Voisin et al. 2013), the z-test based method for elbow detection (Rigamonti et al. 2015), binary classifiers (Yang et al. 2017), a support vector machine (Wu et al. 2017b), as well as a SOM-based approach which calculates the quantization error (Huang et al. 2007).

Prognostics can be carried out using one- or multi-dimensional data. However, many algorithms require a one-dimensional representation of the data in terms of a *health state estimation*. This state estimation is implemented using a health indicator (HI) (Wang 2010, p. 13). The HI is commonly expressed in the interval [0, 1], with the value 0 denoting the healthy condition and value

1 the machine failure. Two types of health indicators are distinguished. The physics HI is derived from one dominant physical signal, which enables a direct measurement of the degradation behavior (Wang et al. 2012, p. 623). The synthesized HI, on the other side, requires a data transformation to a one-dimensional representation. To learn a synthesized health indicator, linear and logistics regression can be applied (e.g., Palau et al. 2020; Bagheri et al. 2015). While logistic regression generates HI values that are less responsive at the beginning and end of the time-series, linear regression has the disadvantage that it allows values outside the interval [0, 1] (Wang et al. 2008, p. 2). Furthermore, direct learning of HI values using an artificial neural network (Riad et al. 2010) or based on the minimal quantization error obtained from a self-organizing map (Huang et al. 2017) can be considered. In order to train algorithms, two strategies have been proposed. Either the method is presented only with healthy and failure data, and values 0 and 1 are assigned respectively, or learning with all degradation data is envisioned (Wang 2010; Ramasso 2014). The latter case requires a function to map the HI values. Here, an exponentially decreasing function is commonly utilized. Another difference depicts whether the health indicator is learned locally or globally with all instances (Ramasso 2014). Lastly, the obtained HI values are adjusted to the interval [0, 1] and smoothed (Riad et al. 2010).

The operation of machines under *dynamic working conditions* is a special case, which often arises in fleet prognostics and is therefore subject to a number of different considerations. With regard to this characteristic, it is assumed that the operating condition influences the sensor readings. Apart from that, there are available condition indicators (Wang 2010). If operating conditions are not explicitly available, the conditions must be extracted, taking into account the individual sensor values. However, in both cases, the number of working regimes needs to be determined and assigned to each time point. For operating condition indicators, this can be realized by clustering approaches (Yu et al. 2012). Given the set of identified regimes, data should be corrected for the effect of various working conditions. Multi-regime normalization presents one approach to handle dynamic working conditions, in which each regime is normalized on its own (e.g., Babu et al. 2016). Another solution is to account for the dynamic operating conditions during the determination of a health indicator. Here, the health values are either defined per regime and fused subsequently or learned with the help of an ANN, which uses the regime and the corresponding sensor values as input (Wang 2010). This procedure can easily be transferred to fleet prognostics. Instead of analyzing the regimes separately, the machines within the same working condition can be prepared together, taking into account the three above-presented options to eliminate this influence.

Numerous algorithms require *labeled data* with associated remaining useful life to all data points. While this is often done straightforwardly by linearly decreasing RUL values, other options include piece-wise linear degradation (Babu et al. 2016; Heimes 2008) as well as power or quadratic functions (Wu et al. 2017b). For improved processing, the negation of the remaining cycles could be considered (Wang et al. 2008; Wang et al. 2012).

Test-Train Split | The last step in data preparation comprises the separation of data into a training and a testing data set. On the one hand, the training data set is used for model building and parameter optimization. For tuning the hyperparameters of ML methods or for model selection, the training dataset is further split to include an additional validation dataset. While the testing data set, on the other hand, presents unseen instances to the model in order to evaluate its performance. In case the training data consists of complete run-to-failure data, data has to be pruned beforehand. This can either be done randomly (Ramasso 2014) or systematically using different intervals (Wu et al. 2017b). To avoid selection bias, cross-validation techniques are frequently applied. Here the data is partitioned iteratively, with one subset representing the testing data and another subset the training data. This is executed multiple times with different combinations of subsets. Cross-validation is applied either only with the training data set (Wu et al. 2017b) or with all instances (Yang et al. 2017). In the first case, a final test is subsequently carried out with the remaining unseen testing data.

5.4.2.2 Fleet Prognostic Method Implementation

The preprocessed training data is subsequently used for developing a prognostic algorithm. The process is highly iterative; thus, returns to the data preprocessing phase might be necessary. In the following, the elaborated fleet approaches are introduced, including their essential steps and possible implementations.

Primarily, elaborated *fleet strategies* are presented. These approaches depict general concepts to handle different fleet characteristics, which are (partly) independent of the specific method. The identified fleet strategies are the local model approach, similarity-based learning approach, correction model approach, Bayesian inference and mixed model approach, as well as ensemble learning approach (cf. Chapter 5.3).

Local Model Approach | In order to capture different trajectory behavior, multiple models can be created for the fleet. Although the creation of local models is very intuitive for heterogeneous fleets, this approach has rarely been discussed in the literature.

Figure 48 illustrates the implementation process for the local model approach.

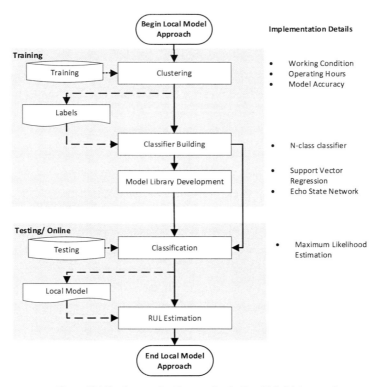

Figure 48 | Implementation Process for the Local Model Approach

Clustering: In the first step, the fleet machines have to be grouped into homogeneous subsets. The grouping depends on the specific application. Options include the working condition (Li et al. 2015), the engine operating hours at the time of failure (Yang et al. 2017), as well as the accuracy with regard to one local model (Peng et al. 2012).

Classifier Building: The most appropriate local model must be determined for online RUL prediction. In the case of the availability of class labels, this is straightforward. Nevertheless, if this information is not available, the corresponding local model needs to be determined analytically. Here, an N-class classifier (Yang et al. 2017) or maximum likelihood estimation (Li et al. 2015) is suggested.

Model Library Development: For each homogeneous sub-fleet, a prediction model has to be developed. In general, all prognostic methods are applicable. Taking into account the identified literature, variants of support vector regressions (Yang et al. 2017; Li et al. 2015), as well as an echo state network (Peng et al. 2012) are applied.

Classification: In order to predict the remaining useful life, a machine must be assigned to one of the sub-fleets by the developed classifier. The obtained label then determines the model used for the RUL prediction.

RUL Estimation: The selected model is subsequently used to estimate the RUL.

Similarity-Based Approach | When there is a large amount of data available, similarity-based approaches are frequently applied. Instead of building one or multiple models, these approaches determine the trajectory similarity of historical data to the current run. For machine j at time t_j, the RUL estimate provided by machine i with regard to its time-to-failure TTF_i is given by:

$$\widehat{RUL} = TTF_i - t_j \tag{5.1}$$

The final RUL combines all estimates considering their similarity. In addition, similarity-based approaches generally require a one-dimensional data representation through dimensionality reduction techniques or a state estimation approach during data preparation. Figure 49 depicts the resulting implementation process for the similarity-based approach. Due to its simplicity, it is the most commonly used approach. Nevertheless, it requires a clear degradation trend and is computational complex during runtime.

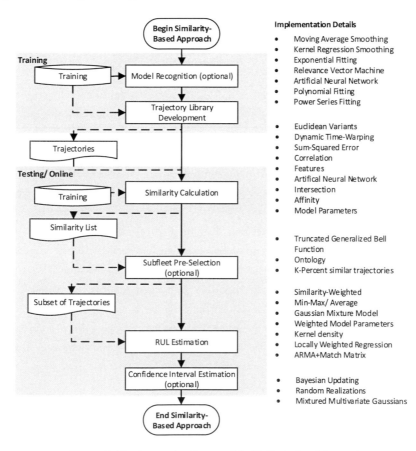

Figure 49 | Implementation Process of the Similarity-Based Approach

Model Recognition/ Trajectory Library Development: The machine to be analyzed must be matched with a degradation library consisting of historical fleet data. To obtain the best possible alignment based on the shape, trajectory abstraction is required to avoid the influence of noise during similarity calculation. This step extends the smoothing operation in the data preparation step. If performed before, it could possibly be skipped. As presented above, methods include moving average with linear interpolation, kernel regression smoothing (Wang et al. 2008; Wang 2010), exponential fitting (e.g., Palau et al. 2020), relevance vector machine (e.g., Chang et al. 2017), polynomial model fitting (Guepie and Lecoeuche 2015), two-term power series fitting as well as artificial neural network fitting (Bektas et al. 2019). In addition, similar degradation trajectories could be grouped beforehand, and trajectories are only defined for each group (Gebraeel et al. 2004; Gebraeel and Lawley 2008).

Subfleet Pre-Selection (optional): While it might be infeasible to consider all trajectories for similarity and RUL estimation, a pre-selection is proposed in some approaches. This can be done by considering only instances in which features are within the support of a truncated generalized bell function (Xue et al. 2008), with regard to an ontology to select a sub-fleet based on similar characteristics (Voisin et al. 2013) or simply by targeting only the k-percent most similar machines (Eker et al. 2014).

Similarity Calculation: There are numerous options to assess the similarity of trajectories, with no measure being used predominantly. On the one hand, these measures are single-point distance measures, such as Euclidean distance (e.g., Yu et al. 2012), including its variants with a time lag and degradation acceleration (Wang 2010), Mahalanobis distance (Bleakie and Djurdjanovic 2013; Liu et al. 2007), dynamic-time-warping (Lam et al. 2014), and the sum-squared error (e.g., Guepie and Lecoeuche 2015). On the other hand, similarity can be assessed considering the correlation between two trajectories, in terms of the Grey correlation degree (Yu et al. 2012) and the Pearson correlation (Lam et al. 2014). Bagheri et al. (2015) describe eleven features for similarity assessment; among them are the similarity in the skewness and kurtosis. Another presented option is the usage of trained ANNs per machine. The error term of the machine under investigation with regard to the trained ANN is considered dissimilar (Gebraeel et al. 2004; Gebraeel and Lawley 2008). While these measures do not impose any restrictions on the input, three further measures are proposed, which require an adjusted input. These are the intersection of polygons, for which an imprecise health indicator is required (Ramasso 2014), the affinity, which uses the variance provided by the relevance vector machine constructed in the trajectory library (Chang et al. 2017), as well as similarity of the fitted model parameters (Palau et al. 2020). In order to handle different initial health conditions, the consideration of the best matching location is proposed (e.g., Bektas et al. 2019).

RUL Estimation: Given the similarity of trajectories, RUL has to be calculated. By far, the most commonly applied are similarity-weighted RUL estimations (e.g., Zio and Di Maio 2010), where the different RUL estimates are combined based on their similarity. Furthermore, a weighted min-max calculation (Ramasso 2014; Wang 2010) and the mean or average value of the k most similar machines (Bektas et al. 2019; Bagheri et al. 2015) are presented. Besides these simple fusion methods, model-based approaches are proposed which incorporate the similarity measures. Examples are weight application to model parameters (Gebraeel et al. 2004; Gebraeel and Lawley 2008) or weighted parameter average of neighboring machines (Palau et al. 2020). In addition, similarity-weighted Gaussian mixture models (Bleakie and Djurdjanovic 2013), kernel density estimation with weights (Wang 2010; Lam et al. 2014), locally weighted regression (Xue et al. 2008), and ARMA models based on a match matrix (Liu et al. 2007) are defined. Due to their nature, the last four model-based approaches already create a failure distribution instead of a pure point estimation.

Confidence Interval Estimation (optional): To quantify the estimated uncertainty, a failure distribution is required. For this purpose, three strategies are identified. These are Bayesian Updating using the RUL estimates of previous cycles (Gebraeel and Lawley 2008), random realizations of the trajectory with regard to the uncertainty coefficient given by the relevance vector machine (Wang et al. 2012; Voisin et al. 2013), as well as mixtures of multivariate Gaussians (Liu et al. 2007).

Correction Model Approach | To learn fleet deviating behavior, the correction model approach comprises two models: a global model that learns the general degradation behavior and a correction model that captures the machine-specific deviation of the global behavior. The latter can be done considering influencing covariates or residuals from the general degradation behavior. For this purpose, two implementation processes are derived for the correction model approach, as depicted in Figure 50 (based on covariates) and Figure 51 (with respect to residuals).

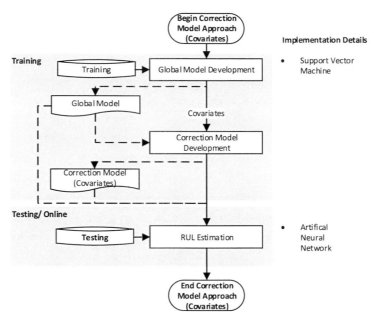

Figure 50 | Implementation Process of the Correction Model Approach (Covariates)

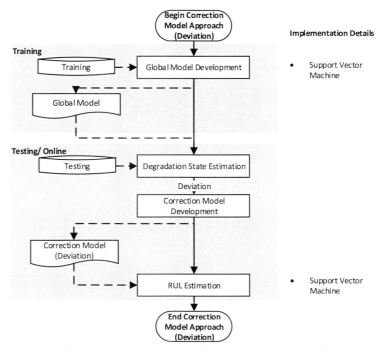

Figure 51 | Implementation Process of the Correction Model Approach (Deviations)

Global Model Development: The general degradation behavior of the fleet is captured in a global model. All prognostic methods are applicable here. Depending on the variant, the model, however, should be able to predict the RUL (using covariates) or future degradation states (using deviations).

Correction Model Development: In order to learn the machine-specific degradation, a correction model can be built subsequently, which incorporates the prediction provided by the global model. Correction models can learn divergences induced by available covariates. This is done during the training phase. For this purpose, ANNs are applied using the general RUL estimate and the covariate as input (Trilla et al. 2018). Another option considers the deviation of the machine's degradation state from the predicted state. The time-series of deviations are used to train the correction model. Here, an SVM has been proposed (Liu and Zio 2016).

RUL Estimation: The prediction is performed for both models. While in the case of covariates, the RUL is depicted by the output of the correction model, for the deviation approach, the RUL estimate of the global model as well as of the correction model are aggregated.

Bayesian Inference & Mixed Models Approach | Bayesian inference is used in fleet prognostics to learn general fleet degradation and to adapt it to machine-specific behavior using the Bayes theorem. It is often used in combination with mixed models, where the parameters are fixed as either deterministic or stochastic. Deterministic parameters are identical for the complete fleet and estimated considering all available data, while stochastic parameters are determined separately for each machine. To infer the value of the stochastic parameter for a machine, Bayesian inference is applied. Figure 52 illustrates the process for the Bayesian inference and mixed model approach.

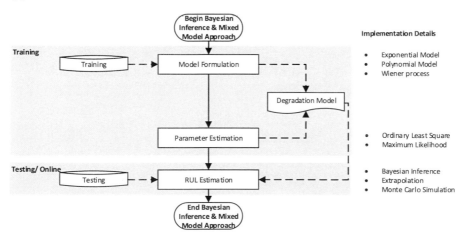

Figure 52 | Implementation Process of the Bayesian Inference and Mixed Model Approach

Model Formulation: A degradation model needs to be defined for general fleet behavior. This is usually done through expert knowledge. Commonly used models are the exponential model (Gebraeel 2006), polynomial model (Zaidan et al. 2013; Zaidan et al. 2015a; Coble and Hines

Wesley 2011), or a Wiener process (Wang et al. 2018). Furthermore, in the case of a mixed model, deterministic and stochastic parameters are determined for the defined model.

Parameter Estimation: Deterministic parameters, as well as prior distributions for the stochastic parameters, have to be defined using the fleet data. This can be done through the ordinary least square (Zaidan et al. 2013; Zaidan et al. 2015a; Coble and Hines Wesley 2011) or the maximum log-likelihood method (Wang et al. 2018). With regard to the parameter variance among machines, the re-definition of parameters can be considered. Parameters with a variance close to zero should be considered deterministic, while high variance indicates stochastic behavior among machines (Coble and Hines Wesley 2011).

RUL Estimation: Bayesian inference is performed to obtain an estimate for the stochastic parameters. The model is extrapolated to a pre-defined threshold. A failure time distribution can then be calculated analytically. Due to its complexity, the Monte Carlo simulation is performed for a Bayesian hierarchical model with two-level parameter dependencies (Zaidan et al. 2015a).

Ensemble Learning Approach | As a meta-approach, ensemble learning combines various methods and fuses the results afterward to increase accuracy. For this, it is necessary to build a model library in which every method is estimating the remaining useful life. This is done to capture the shortcomings of each method. All estimations are merged to get a final remaining useful life. For ensemble learning, the fleet characteristics can either be taken into account in the individual methods or as part of the fusion strategy. Figure 53 presents the resulting implementation process for the ensemble learning approach.

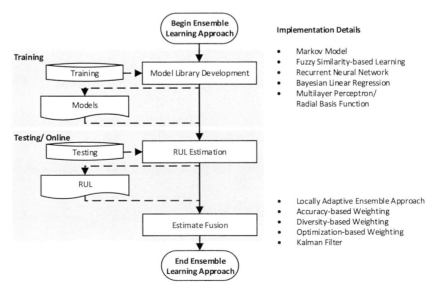

Figure 53 | Implementation Process of the Ensemble Learning Approach

Model Library Development: There are countless possibilities to build up the model library. Besides using multiple methods, different method parameterizations or a combination of both is possible. Ensembles built for fleet prognostics have all created model libraries made up of other fleet prognostic approaches presented above. In particular, these are similarity-based approaches, ANN approaches, Markov model, and a Bayesian Updating approach.

RUL Estimation: The developed models are used to estimate the remaining useful life.

Estimate Fusion: The various estimates are then merged to produce a final RUL. Multiple options are presented in fleet prognostics as a fusion strategy. The locally adaptive approach (Al-Dahidi et al. 2017) assigns a weight to each model based on the accuracy of a similar pattern from the test data set. In Hu et al. (2012), three strategies are presented, namely the accuracy-based approach (weights based on the accuracy of test data set), diversity-based approach (weights to increase robustness via the diversity of estimates) as well as the optimization-based approach (optimizes weighs to achieve accuracy and robustness simultaneously). As another option, a Kalman filter approach is proposed, in which the multiple RUL estimates are considered as observations, and a combined posterior is estimated (Peel 2008; Lim et al. 2014).

Besides the presented fleet strategies, there are various methods that possess *intrinsic fleet capabilities*. These methods can either directly incorporate influences of covariates (ANN, PHM), model different degradation states (Markov models), or capture correlations between the machine's degradation trajectories via an additional covariance (GPR). These methods, however, need to address the fleet characteristics adequately and are thus explained and visualized in the following.

Artificial Neural Networks | Due to their ability to learn non-linear structures from data, ANNs are suitable for application within fleet prognostics. In the case of explicit covariates in terms of, e.g., working conditions or machine type, this information should be provided to the network as a feature. Nevertheless, in general, artificial neural networks are able to learn different structures and dissimilarities from the data. Figure 54 depicts the process for the development of an artificial neural network for fleet prognostics.

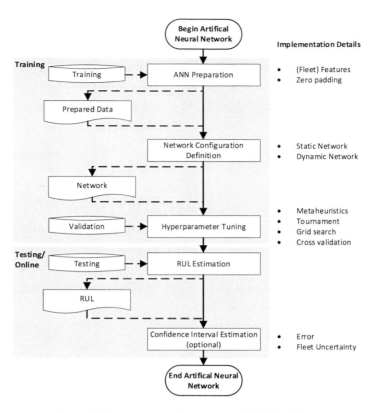

Figure 54 | Implementation Process for Artificial Neural Networks

ANN Preparation: Artificial neural networks require the definition of an output label during training. For this purpose, data labeling, as described in Chapter 5.4.2.1, is required. In addition, fleet features should be extracted from the data. These include, among others, the current operating mode (Wu et al. 2017b; Heimes 2008), the operating history (Babu et al. 2016; Wu et al. 2017b; Peel 2008; Lim et al. 2014) as well the difference in signal value with regard to the same operating condition (Wu et al. 2017b). In general, available information on the dissimilarity between the machines should be encoded as feature input to the network. Furthermore, due to the varying length of time-series, a zero-padding at the beginning is suggested in combination with a masked layer to identify non-informative inputs (Wu et al. 2017b).

Network Configuration Definition: There are various network structures for ANN. For prognostics, these can be distinguished into static networks and dynamic networks (Sikorska et al. 2011, p. 1827). In static networks, inputs to a layer are only dependent on the output of the previous layer. Identified static network structures used in fleet applications are multi-layer perceptions, radial basis functions (Peel 2008; Lim et al. 2014), as well as the convolutional neural network, which exhibits a deep architecture (Babu et al. 2016). On the other side, dynamic networks are recurrently using outputs from previous iterations to include historical information in the learning process. Besides simple recurrent neural networks (Heimes 2008), echo state networks with sparse connected hidden layers (Rigamonti et al. 2015) and long-term short-memory networks (Wu et al. 2017b) are identified as options.

Hyperparameter Tuning: Hyperparameters are ANN network parameters that must be specified prior to network learning. These include, among others, the number of hidden layers as well as the number of neurons per layer. There are options to search for suitable values experimentally. This is realized on the validation data set. For this, the range of possible parameter values is specified, and a thorough search is conducted. In the identified publications, the search is performed using the metaheuristics genetic algorithm (Heimes 2008) or differential evolution (Rigamonti et al. 2015), grid search (Wu et al. 2017b), cross-validation (Babu et al. 2016; Wu et al. 2017b), and a tournament-style heuristics (Peel 2008; Lim et al. 2014).

RUL Estimation: Using the learned neural network, RUL estimation is straightforward. The data is prepared as described in ANN preparation, and the resulting input features are fed to the neural network. To cope with variations in the output values through time, RUL estimates can be smoothed.

Confidence Interval Estimation (optional): In general, ANN models provide point estimates of the remaining useful life. To quantify the uncertainty, the distribution of the deviation (error during testing) can, for example, be transferred from the fleet (Cristaldi et al. 2016).

Markov Models | The ability to learn different states resulting from heterogeneous degradation behavior enables Markov models to be used for fleet prognostics. Depending on the experienced operating conditions, alternative degradation paths can be formed by the identified states. In order to learn different states of degradation adequately, the presence of influential covariates is adjuvant. Figure 55 represents the implementation process for Markov models.

State Identification: Markov models require the identification of distinct states of degradation. In a data-driven approach, these states are determined from historical data. An unsupervised ensemble clustering approach considering different operating conditions has been proposed by Al-Dahidi et al. (2016). Here, two clusterings are performed, one representing the signal behavior and one other the operating conditions. Through a co-association matrix, a consensus clustering is defined and evaluated using the Silhouette coefficient. This leads to a set of defined states considering both, the degradation behavior and the operating condition.

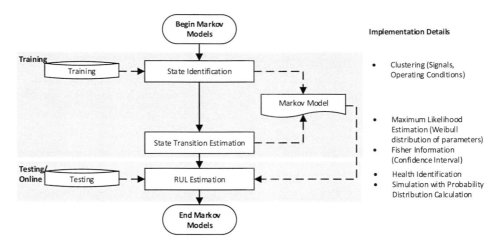

Figure 55 | Implementation Process for Markov Models

State Transition Estimation: Thereafter, the state transition probability needs to be defined. For this purpose, the distribution and its parameters are optimized. One option is the Weibull distribution, for which the parameters can be obtained through Maximum Likelihood Estimation (Al-Dahidi et al. 2016). Given the estimates, the Fisher Information Matrix is used to gain the corresponding confidence intervals (Al-Dahidi et al. 2016).

RUL Estimation: The obtained model can be used for RUL estimation. For this purpose, the current health state of the machine needs to be identified, considering the condition measures and covariates. Thereafter, various simulation runs of the model are performed by sampling the sojourn times for the remaining states. A probability distribution can be calculated with respect to the RUL estimates of the simulation runs (Al-Dahidi et al. 2016).

Gaussian Process Regression | To model the different machine degradations simultaneously and thus, exploit the similarity between the trajectories, Gaussian multi-output process regressions (MO-GPR) are applied. As an extension of single-output Gaussian process regressions, they incorporate an additional covariance to specify the correlation between the degradation trends of individual machines. The extended covariance κ_{MO-GPR} is then defined as the product of the covariance κ_{yy} between different cycles within one machine and the covariance κ_c between different degradation trends. For this purpose, MO-GPR depicts a fleet prognostic method, which can detect differences among machines without resorting to covariates. In Figure 56, the identified implementation process for Gaussian Process Regression is presented.

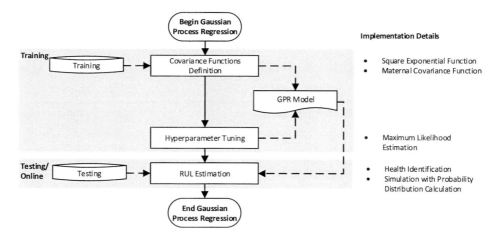

Figure 56 | Implementation Process for Gaussian Process Regression

Covariance Functions Definition: Both covariance functions need to be determined. For the covariance function κ_{yy}, the square exponential and maternal covariance functions are frequently employed (Duong et al. 2018). The covariance κ_c requires to be positive semi-definite. Thus, the free-form parametrization using Cholesky decomposition is proposed (Duong et al. 2018).

Hyperparameter Estimation: Hyperparameters for both covariances as well as the noise variance are to be estimated. This can be done by maximizing the likelihood (Duong et al. 2018).

RUL Estimation: Given the prior distribution of the training data, the posterior prediction mean and variance can be estimated, resorting to the Bayesian paradigm (Duong et al. 2018).

5.4.2.3 Performance Evaluation

Performance evaluation depicts the concluding step in the fleet prognostic algorithm implementation phase. Different performance metrics are used either to perform parameter optimization or to assess the overall quality of the solution. Depending on the result, further iterations are required.

In general, three functional groups of prognostic metrics can be specified, namely algorithm performance, computational performance, and cost-benefit (Saxena et al. 2008a, p. 6). With respect to the target user and application, different metrics can be used. As this section focuses on the evaluation of the algorithm performance, only these identified metrics are further elaborated. Furthermore, the identified metrics do not specifically target fleet data but are generally applicable to prognostics.[31]

Within the group of algorithm performance metrics, three major classes can be distinguished with different objectives and emphasis. Most of the identified metrics belong to the first class, which denotes *accuracy-based metrics* (Saxena et al. 2008a, pp. 8 f.). These evaluate the quality of a prognosis in terms of the error between the ground truth and the estimates. Known representatives of this group are the mean absolute error (MAE), mean squared error (MSE), and the mean absolute percentage error (MAPE), including their variants. These metrics are used frequently. In addition, the bias metric and the anomaly correlation coefficient are each considered once. While these metrics directly measure the deviation of the RUL prediction in terms of an error metric, another option results from defining an acceptable range of deviation and classifying the predictions accordingly. These metrics originate from classification tasks and include the false-positive and false-negative rates.

The second class, *precision-based metrics* (Saxena et al. 2008a, pp. 9 f.), evaluates the distribution of the estimates. Within this group, two metrics are used by fleet prognostic algorithms, namely the sample standard deviation and the mean and median absolute deviation from the sample median. As these metrics do not provide any information on the reliability of an algorithm with regard to the ground truth, they are not frequently applied and only used in combination with metrics from the accuracy-based class.

In addition to these traditional performance metrics derived from other forecasting and classification applications, performance evaluation is considered specifically in the context of prognostic algorithms. To capture the shortcomings of general performance assessment, several metrics are developed within the field of PHM. These metrics include the performance improvement over time as more information is available, the need for a reliable prognosis near the end of life as well as the emphasis on early prediction compared to late predictions. Metrics in the third class, the *prognostics performance metrics*, are therefore developed specifically for the given domain and explained in the following.

The first prognostics metric is the *prediction horizon (PH)*. The metric prediction horizon reflects the period for which an appropriate RUL is estimated. It is defined as the difference between the time at index i_α where the prediction first passes the α-bound criteria and the failure time (t_{EOL}) (Saxena et al. 2010a, p. 14; Wang 2010, p. 84):

---

[31] The summary of the identified metrics is shown in Table 39 in Appendix H.

$$PH = t_{EOL} - t_{i_\alpha} \tag{5.2}$$

In order to derive index i_α, at each prediction time stamp t_i from the starting time of the prognostics t_s until the end of lifetime t_{EOL}, the α-bound criteria is evaluated by assessing whether the point estimation $\hat{Y}_i$ falls within the α-bound of the true RUL (Y_i).

$$t_{i_\alpha} = min\{t_i \mid t_i \in [t_s, t_{EOL}], Y_i - \alpha * t_{EOL} \leq \hat{Y}_i \leq Y_i + \alpha * t_{EOL}\} \tag{5.3}$$

The second metric, the *rate of acceptable predictions (RAP)*, calculates the rate of estimates, which fall in a region of acceptable prediction errors. This region defines a conical area. This metric depicts an extension of the $\alpha - \lambda$ accuracy metrics introduced by Saxena et al. (2010a, pp. 14 f.). The RAP is calculated as follows (Wang 2010, pp. 85 f.):

$$RAP = mean(\{\delta_i | t_H \leq t_i \leq t_{EOL}\}) \tag{5.4}$$

$$\delta_i = \begin{cases} 1 & if \ (1-\alpha)Y_i \leq \hat{Y}_i \leq (1+\alpha)Y_i \\ 0 & otherweise \end{cases} \tag{5.5}$$

Where t_H denotes the time of the first appropriate prediction, as calculated by the prediction horizon.

The third metric, the *relative accuracy (RA)*, adapts the MAPE metric for all $t_i \geq t_H$ (Saxena et al. 2010a, p. 15; Wang 2010, p. 87). Instead of all predictions, it considers only later RUL estimations (Saxena et al. 2010a, p. 13). The threshold t_H can be defined with respect to the prediction horizon (Wang 2010, p. 85).

$$RA = 1 - Mean(\left\{ \left| \frac{Y_i - \hat{Y}_i}{Y_i} \right| \Big| t_H \leq t_i \leq t_{EOL} \right\}) \tag{5.6}$$

The fourth metric, *convergence (CG)*, evaluates the speed of performance improvement with new data (Saxena et al. 2010a, p. 16; Wang 2010, pp. 87 f.).

$$CG = \left(\frac{\frac{1}{2}\sum_{i=P}^{EoL}(t_{i+1}^2 - t_i^2)M_i}{\sum_{i=P}^{EoL}(t_{i+1} - t_i)M_i} - t_p \right) * \frac{1}{t_{EOL} - t_p} \tag{5.7}$$

Where M_i could be any performance metric. In the simplest way, the absolute error $M_i = Y_i - \hat{Y}_i$ is applied.

Lastly, the *PHM score of* data competition is another prognostics performance metric specifically designed for prognostic requirements. The underlying assumption is that late predictions should be penalized stronger than early predictions, as the former would lead to machine failure (Saxena et al. 2008b, p. 7). The PHM score is, therefore, defined as follows:

$$PHM_{score} = \sum_{i=1}^{n} S_i \tag{5.8}$$

$$S_i = \begin{cases} e^{-\frac{d_i}{13}} - 1 & d_i < 0 \\ e^{\frac{d_i}{10}} - 1 & d_i \geq 0 \end{cases} \tag{5.9}$$

where d_i depicts the prediction error calculated as the estimated RUL minus the true RUL ($\hat{Y}_i - Y_i$).

In order to evaluate multiple metrics simultaneously, in particular for parameter optimization, a combined metric with a weighted sum of the individual metrics is presented in Wang (2010, p. 88), called the *performance score.*

All presented performance metrics provide a different view on the prognostic quality of the developed algorithm and can be seen as complementary to each other. The metric selection should thus be aligned with the prognostic objective of the targeted application. In general, the accuracy-based metrics are used predominantly in the literature and enable a comparison to existing methods. Consequently, these metrics form a good starting point for assessing the prognostic performance. However, it is advisable to supplement these metrics with the prognostics performance metrics, as they have been developed specifically for the application context taking into account the key prognostic characteristics. The prognostic horizon metric indicates when the RUL estimate can be trusted for maintenance scheduling. The prognostic horizon should thus be sufficiently large to allow maintenance to be scheduled and spare parts to be ordered before the machine fails. The metrics rate of acceptable prediction, relative accuracy and convergence, further specify the prognostic quality towards the end of the remaining useful life. In case a very precise prognostic estimate is required close to failure, these metrics should be considered and optimized for the targeted application. Lastly, the PHM score aims to prevent late predictions by penalizing them. This metric is particularly relevant for critical machines. In general, however, it is not possible to optimize all metrics simultaneously. Thus, one or two metrics should be selected for optimization while defining minimum thresholds for the other relevant metrics.

Table 23 provides an overview of the key characteristics, as well as the identified advantages and disadvantages in order to support the metric selection.

Table 23 | Evaluation of Performance Metrics

Metrics	Key Characteristics	Advantage	Disadvantage
Accuracy-Based Metrics	General accuracy (error between the ground truth and the estimates)	• General accuracy • Enable the comparability to other approaches in the literature	• Do not incorporate prognostic requirements
Prediction-Based Metrics	Distribution of estimates	• Provides further information on the distribution of estimates	• No accuracy considered • Should be applied together with other metrics
Prognostic Horizon	Time horizon of acceptable predictions	• Provides a reliable lead time for maintenance scheduling	• No accuracy considered
Rate of Acceptable Prediction Relative Accuracy Convergence	Prediction quality near the end of the RUL	• Enables accurate RUL estimates close to machine failure	• Disrespect the quality of predictions with long RUL
PHM Score	Early predictions are superior to late predictions	• Aim to minimize machine breakdowns	• Can lead to too early maintenance of machines

5.5 Evaluation of the Algorithm Development Method

In the following, the algorithm development method is evaluated by means of multiple instantiations. In particular, the applicability of the method is targeted. For this purpose, the data sets of turbofan engine degradation as presented above (cf. Chapter 4.5.1) are considered, and the identified fleet types are drawn upon.

The evaluation of the algorithm development method is organized in consideration of the two phases: in Chapter 5.5.1, the selection of fleet approaches for the four data sets is discussed, taking into account two different scenarios. This is followed in Chapter 5.5.2 by the implementation of the selected fleet approaches. Lastly, the results of the evaluation are discussed in Chapter 5.5.3.

5.5.1 Fleet Approach Selection

Along with the fleet type and the data characteristics, the selection of the methods is based on a variety of criteria that result from the specific application context. However, this information is not available for the targeted data sets. Therefore, two very different scenarios are defined for the evaluation. The first scenario targets a company at the initial stage of predictive maintenance implementation. The emphasis is on the initial understanding and insights of the data, as well as a feasibility analysis for fleet prognostics. Therefore, a simple and straightforward implementation is pursued, which strives for a high degree of transparency. In contrast, the second scenario addresses a company for which accuracy of the remaining useful life prediction is of utmost importance. The company would like to base its maintenance decisions on the prediction.

Regardless of the scenarios, several characteristics which are relevant for fleet approach selection, can be identified based on the results of the characterization methods (cf. Chapter 4.5). From the fleet description, it can be inferred that the fleets are large, and the data is multivariate, non-stationary, and includes complete run-to-failure data. Furthermore, to the data set FD001 and FD002, a homogeneous fleet prognostic type is assigned, while for FD003 and FD004, a heterogeneous fleet prognostic type is identified.

For scenario one, two further criteria are derived, namely high transparency and low parameter handling. Later is required to ease the implementation. For the homogeneous fleets, these characteristics are best reflected by the similarity-based approach, while for the heterogeneous fleets, the local model approach should be targeted in view of the fleet approach selection process. The latter requires the additional selection of a prognostic method. With regard to the application characteristics, simple trend regression, stochastic processes, and Markov models are eligible. As the last two (partly) require additional expert knowledge as well as are very restrictive in terms of their system assumption, the simple trend regression is targeted. Moreover, for scenario two, the key characteristic depicts high accuracy. This leads to the selection of artificial neural networks for all four data sets. Table 24 summarizes the identified fleet approaches.

Table 24 | Overview of Selected Fleet Approaches

	Key Characteristics	Homogeneous Fleet (FD001/ FD002)	Heterogeneous Fleet (FD003/ FD004)
Scenario 1 (Simplicity)	High Transparency Low Parameter Handling	Similarity-Based Approach	Local Model Approach using Simple Trend Regression
Scenario 2 (Accuracy)	High Accuracy	Artificial Neural Networks	

5.5.2 Fleet Prognostic Implementation

For the two scenarios, the implementation for the four data sets is highlighted in the following.[32] The implementation is realized taking into account the detailed steps and design options presented within the developed method. In addition, the results are briefly evaluated relative to a global model enabling a conclusion to be drawn about the accommodation of fleet characteristics.

First, scenario one is targeted. For this, the similarity-based approach is presented in Chapter 5.5.2.1, and the local model approach in Chapter 5.5.2.2. For scenario two, the same approach is chosen for both fleet types; hence, the implementation is described jointly in Chapter 5.5.2.3.

---

[32] The specific steps are conducted and implemented in Python Version 3.7 using several additional python libraries. A list of all packages is provided in Table 40 in Appendix I.

5.5.2.1 Similarity-Based Approach Algorithm Development Process

The similarity-based approach considers the two data sets attributed to the homogeneous fleet prognostic type (FD001 and FD002). For this purpose, a similarity-based algorithm development process is deduced from the method. Figure 57 highlights the phases, steps as well as detailed design options. A detailed description of the steps is given below.

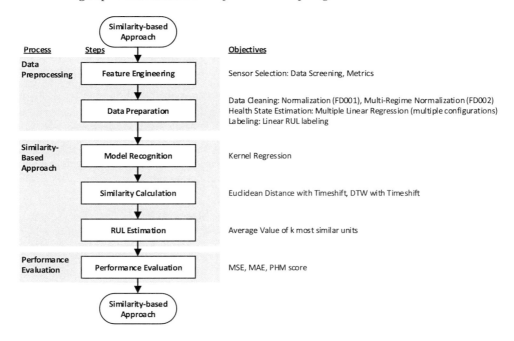

Figure 57 | Similarity-Based Algorithm Development Process

Feature Engineering | Since both data sets consist of 21 sensors, not all of which are relevant and have underlying trends, the first step involves reducing the data to the most relevant sensors. This is done twofold. On the one hand, simple data screening is performed. For this purpose, sensors are excluded with no trend, as identified in Table 15 in Chapter 4.5.2. This results in a subset of 14 sensors. Besides, sensors are selected which satisfy the three metrics of monotonicity, trendability, and prognosability. This approach yields ten sensors.

Data Preparation | With respect to the identified sensor subset, data is further cleaned and processed. In particular, normalization (for FD001) and multi-regime normalization (FD002) are performed (analogously to Chapter 4.5.2). As the similarity-based approach requires univariate data, a health state estimation is performed by means of multiple linear regression. This approach is chosen due to its higher sensitivity towards the end of the remaining useful life. In addition, the disadvantage of potential values outside the range of [0,1] is not applicable in this case since the RUL is calculated independently of the health status with respect to the trajectory similarity. In accordance to Wang et al. (2008), the linear regression model is trained on the first

177

n values, which are assigned a healthy condition ($HI = 0$) and the last n values labeled as faulty ($HI = 1$). For n, three different values ($n = \{5, 10, 15\}$) are investigated. Lastly, linear RUL labels are assigned to the training data sets.

Model Recognition | In order to match the univariate time-series best, the trajectories should be smoothed using a suitable model. In agreement with Wang (2010, pp. 45 f.), kernel regression is selected due to its good performance.

Similarity Calculation | For the assessment of similarity, the measures Euclidean distance and dynamic time warping are selected, as they have proven to be robust within the conducted factor impact analysis (cf. Chapter 4.5.2). In addition, to overcome the problem of different initial health conditions, the best match of the time-series is identified through time-shifting.

RUL Estimation | The remaining useful life is calculated between failure cycle of the training machine j (max_cycle_j) and the last recorded cycle of the test machine i (max_cycle_i) as well as the time shift of the best matching location ($time\_shift_{i,j}$) as follows:

$$rul_{i,j} = max\_cycle_j - \text{max}\_cycle_i - time\_shift_{i,j}$$

(5.10)

Figure 58 visualizes the similarity-based approach with regard to one training and test degradation trajectory. In order to combine the individual RUL estimations, the final remaining useful life is calculated as the mean value of the k most similar trajectories (cf. Bektas et al. 2019). For this purpose, three values for k are tested ($k = \{5,10,15\}$).

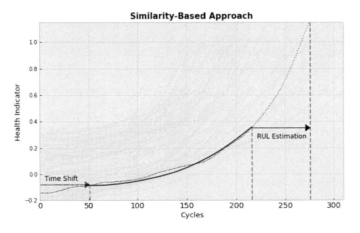

Figure 58 | Similarity-Based Approach Visualization

Performance Evaluation | For each combination of the described design options, the performance is evaluated with regard to the metrics MSE, MAE, and the PHM score. The best-performing configuration is selected thereafter.

In addition, results are compared to a global model. For the global model, simple trend regression is chosen due to their similar characteristics in the prognostic method assessment (cf. Table 22). To enable comparison, the same data preprocessing steps is applied. In order to account for different initial wear, the model is trained on negative RUL values ($RUL = -RUL$). To model the degradation, two data representation models are selected, namely the exponential and the polynomial model. As the exponential model outperforms the polynomial model on the metrics MAD and MSE, it is taken for the final comparison. The RUL estimates are predicted with respect to the inverse exponential function (logarithm function) of the last measured time point.[33] The final results are shown in Table 25.

Table 25 | Results Similarity-Based Approach

	Best Design Option	MAE		MSE		PHM Score	
		Fleet	Global	Fleet	Global	Fleet	Global
FD001	Sensor Selection: Data Screening n = 15 Distance Measure: DTW K = 15	12.38	21.79	302.15	911.33	687.48	3.20E+4
FD002	Sensor Selection: Data Screening n = 5 Distance Measure: Euclidean K = 15	15.19	28.2	484.29	1774.85	7114.79	6.81E+7

Despite the simple approach, good results were obtained for the similarity-based approach. In particular, it can be recognized that the approach can handle the fleet characteristics significantly better than the benchmarked global model.

5.5.2.2 Local Model Approach Algorithm Development Process

For the heterogeneous fleets (FD003 and FD004) within scenario one, a local model approach is selected. Following the fleet prognostic implementation process, a local model algorithm development process is devised.

Figure 59 depicts the resulting process by highlighting the three phases, the corresponding steps as well as the targeted objectives.

---

[33] The configuration of the global model is described in detail in the following sub-chapter as part of the step model library development.

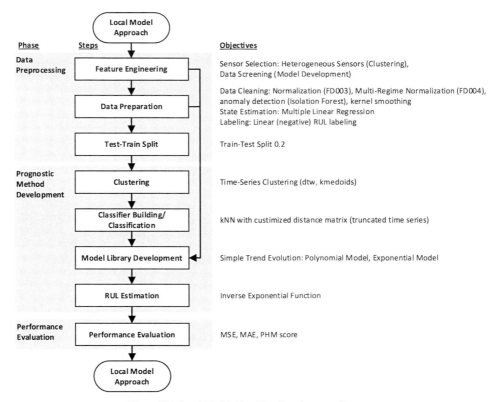

Figure 59 | Local Model Algorithm Development Process

Feature Engineering | For the data preprocessing phase, a distinction must be made between the clustering and classifier building step and the model library development step. While the former requires the subset of only heterogeneous sensors, the model development is performed on all sensors which exhibit a trending behavior. This is deemed useful because the additional sensors provide further information about the degradation behavior, which could be of value for the remaining useful life prediction.

Data Preparation | Within data preparation, first, data is normalized. This is done through normalization (FD003) and multi-regime normalization (FD004) in order to handle the impacts of different operating conditions. For clustering and classifier building, the data is prepared, as described in Chapter 4.5.2. This comprises anomaly detection of distorted sensor values as well as kernel smoothing. In addition, for classifier building, the training data sets are split (80% training and 20% testing) in order to evaluate the performance and be able to select the most suitable classifier. For the step model library development, the data sets are reduced to a univariate representation. Analogously to Chapter 5.5.2.1, a multiple linear regression model is computed for each cluster. Furthermore, the data sets are labeled by negative linearly decreasing RUL values in order to account for initial wear.

Clustering | In the first step of the prognostic method development, homogeneous sub-fleets have to be identified. This is achieved on the basis of the time-series clustering approach, as presented in Chapter 4.4.2. For this purpose, sensors are selected, which have been identified to exhibit heterogeneous behavior. Furthermore, for time-series clustering, dynamic time warping is considered as the distance measure due to its robustness and capability to handle time-series of unequal length. In addition, k-medoids is chosen as the clustering method given its fast-processing time and high quality.

Classifier Building/ Classification | After the data sets are labeled within the clustering step, a classifier is constructed to identify the matching clusters for the test data sets. For this purpose, the k-nearest neighbors (kNN) classifier is chosen with a customized distance matrix as this constitutes the prevalent method within time-series classification (Abanda et al. 2019, p. 378). The distance matrix is computed using the same configurations. In order to find the best classifier, different sensor combinations are tested. For this purpose, the training data sets are split using a test-train-split of 0.2. All sensor combinations result in 100% accuracy on the test data sets. Thus, to better reflect the characteristics of the test data set, the time-series in the training data sets are truncated by 25, 50, 75, 100, and 125 values, and the performance of the classifier is tested. The best classifier on the truncated data sets is used for calculating the labels of the test data sets.

Model Library Development | The training data set is split with regard to the identified cluster labels. For each cluster, a simple trend regression model is targeted. Subsequently, both a polynomial and exponential model is implemented (in accordance with Coble (2010) and Wang (2010)) with respect to the following equations:

$$HI_t^{exp} = a * \exp(b * x_t + c) + d \tag{5.11}$$

$$HI_t^{poly} = a * x_t{}^2 + b * x_t + c \tag{5.12}$$

In the two equations. $a, b, c,$ and d represent parameters, which must be fitted to the data. In addition, x depicts the negated remaining useful life at time t and HI_t the health indicator at time t. The models are evaluated using the two metrics MAE and MSE. It is shown that the exponential model outperforms the polynomial model for all clusters. Figure 60 visualizes both models for the first cluster of FD003.

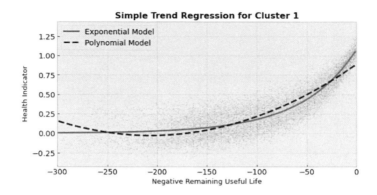

Figure 60 | Simple Trend Regression for Cluster 1

RUL Estimation | For RUL estimation, the inverse exponential function is calculated with respect to the fitted exponential model. Given the last health indicator $HI_{current\ time}$, the remaining useful life can thus be computed following the equation:

$$x_t^{exp} = (\ln\left(\frac{HI_t - d}{a}\right) - c) / b \qquad (5.13)$$

As the natural logarithm $\ln(x)$ is undefined for negative values, in these cases, the RUL is set to $x_t^{exp} = -130$ in accordance with Heimes (2008, p. 4). This point is defined as the mean turn point between the healthy state and the onset of machine degradation. As $x <= 0$ most likely relates to a small health indicator, this is a reasonable assumption, even though it might underestimate the remaining useful life.

Performance Evaluation | The performance of the algorithm is evaluated by means of the three metrics MSE, MAE, and the PHM score.

Finally, the results of the local model approach are compared to a global model, which is trained on the data sets irrespectively of the two identified clusters. To enable comparison, the same data preprocessing steps are followed, and the same method is taken. Table 26 provides an overview of the achieved results. It can be recognized that the local model approach improves the overall accuracy relative to the global model with respect to both clusters.

Table 26 | Results Local Model Approach

	Fleet Cluster	MAE		MSE		PHM Score	
		Fleet	Global	Fleet	Global	Fleet	Global
FD003	Cluster 1	21.68	27.49	1373.82	2167.47	4.01E+6	1.19E+11
	Cluster 2	23.41		1105.77		3.89E+4	
FD004	Cluster 1	26.0	35.23	1372.46	2705.71	2.10E+4	1.88E+10
	Cluster 2	29.24		1917.19		1.08E+10	

The local model approach provids an appropriate and simple way to address the heterogeneity of the fleet by considering the two sub-fleets separately. The two homogeneous fleets, however, are targeted by a global model within the model library development step. To improve the accuracy, a fleet strategy for homogeneous fleets could be considered.

5.5.2.3 ANN Algorithm Development Process

For the second scenario, artificial neural networks are considered for both fleet prognostic types because of their inherent ability to learn complex patterns. In addition, the method is claimed to achieve a high degree of accuracy. In accordance with the method, an algorithm development process is devised, as illustrated in Figure 61.

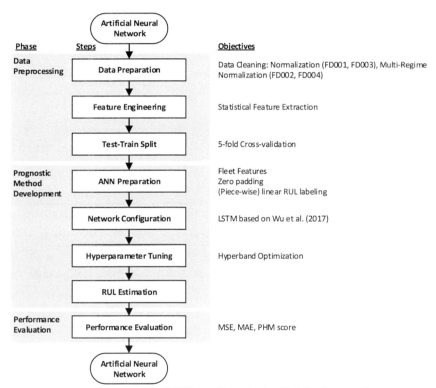

Figure 61 | Artificial Neural Network Algorithm Development Process

The implementation of an artificial neural network algorithm depends on the selected network structure. For the evaluation, a vanilla long short-term memory (LSTM) neural network is considered as proposed by Wu et al. (2017b). This variant of a recurrent neural network has become the state-of-the-art model within dynamic neural networks, as it is able to learn long-term dependencies (Wu et al. 2017b, p. 1). For this purpose, LSTM networks are augmented with memory cells, which can store information over a long period of time (LeCun et al. 2015, p. 442). The information in the memory cell is controlled by three types of gates: the input, forget and output gate (Goodfellow et al. 2016, pp. 406 f.). The forget gate identifies the information which should be removed, the input gate updates the cell state, and the output gate defines the next hidden state (Wu et al. 2017b, p. 3). Both the cell state and the hidden state is forwarded to the next time step.

Data preparation | Initially, the data sets are prepared. Analogous to the other approaches, normalization is performed for FD001 and FD003 and multi-regime normalization for FD002 and FD004. The latter is required due to the six operating conditions, which affect the sensor values.

Feature Engineering | Feature engineering is key for the development of an ANN. The identified features are input to the network. Various statistical features are extracted to increase data dimensionality. For each sensor, eight statistics are calculated (root mean square, variance, mean, minimum, maximum, skewness, kurtosis, peak-to-peak) as proposed by Zhao et al. (2017). To account for local and global features, both a sliding window of size 5 and an expanding window is chosen. Lagged features are added last. These include the three past time steps as well as the difference to the previous step.

Test-Train Split | To tune parameters, 5-Fold cross-validation is selected as recommended by Babu et al. (2016).

ANN Preparation | In addition to feature engineering, further data preparation steps are required for the implementation of ANNs. In particular, for fleet prognostic algorithm implementations, different fleet features are identified as part of the algorithm development method. Since these fleet features are geared to different operating conditions, they have only been generated for FD002 and FD004. In particular, two types of features are proposed; namely, the number of cycles spend within the different operating conditions as well as the feature change since the last cycle within the same operating condition (Wu et al. 2017b, p. 3). For this purpose, the different operating condition clusters are taken, as identified in Chapter 4.5.2, and features are derived accordingly. Furthermore, zero-padding is performed at the beginning of the time-series in order to get time-series of equal length for training. In addition, a masked layer is added to identify these non-informative inputs. Lastly, artificial neural networks require an output label during training. Two alternatives are selected for RUL labeling, namely the simple linear function as well as the piece-wise linear function as proposed by Heimes (2008) and Wu et al. (2017b). The latter assumes that degradation becomes visible in the data at approximately 130 cycles prior to the failure.

Network Configuration | First, a baseline implementation is targeted. For this purpose, the network configurations, as described by Wu et al. (2017b), are adopted. For the vanilla LSTM, these are the following configurations: number of hidden layers: 256, recurrent activation function: sigmoid function, activation function: hyperbolic tangent (tanh), dropout rate: 0.5, unit forget bias: 1, optimizer: adam and loss function: MSE. In addition, the early stopping criterion is set to 5. For the remaining parameters, the default values by Keras are kept. Figure 62 depicts a visualization of the resulting LSTM network.

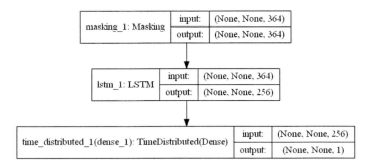

Figure 62 | Vanilla LSTM Network Structure (FD001)

With regard to this baseline configuration of the LSTM neural network, both the RUL labeling function and the fleet features are evaluated. Table 27 depicts the results. For each run, the cross-fold MAE and MSE (mean value of all five runs) are calculated. With regard to the RUL labeling function, it can be seen that the piece-wise linear function (depicted as 130) significantly improves both metrics over the linear function. In contrast, results for the fleet features (labeled as fleet) are inconclusive. While it could be argued that they have a minor impact, no definitive statement can be drawn. In particular, the effects of multiple operating conditions are already eliminated by the multi-regime normalization. Nevertheless, the fleet features (for FD002 and FD004), as well as the piece-wise linear RUL function, are considered within hyperparameter tuning as they yield the best results.

Table 27 | Training Results of the Vanilla LSTM

		Cross-Fold MAE	Cross-Fold MSE
FD001	FD001_130	**11.10***	**118.23***
	FD001_linear	22.47	688.29
FD002	FD002_130_fleet	**9.47***	**99.07***
	FD002_linear_fleet	23.91	645.72
	FD002_130	10.33	116.49
	FD002_linear	23.87	647.15
FD003	FD003_130	**9.56***	**85.17***
	FD003_linear	36.35	1559.43
FD004	FD004_130_fleet	11.85	**117.11***
	FD004_linear _fleet	37.49	1229.68
	FD004_130	**11.32***	119.21
	FD004_linear	35.61	1162.99

Legend: *best results for each data set during model training (marked in bold)

Hyperparameter Optimization | In order to tune the network further, hyperparameter optimization is performed. Within an empirical study, it is identified that for LSTMs, the most influential parameters are the learning rate as well as the hidden layer size (Greff et al. 2017, p. 2228). In addition, as the learning rate and the batch size interfere with each other, they should be tuned jointly (Breuel 2015, pp. 5 f.). Consequently, optimization is carried out for these three hyperparameters by means of the hyperband method. Hyperband, developed by Li et al. (2018a), is a random configuration sampling approach to hyperparameter optimization that aims to fasten optimization by disregarding configurations that perform poorly in the initial runs. For the three parameters, the search space for hyperband is defined as follows:

- learning rate: sampled uniformly between 0 and 1
- batch size: sampled from the values 5, 10, 20, 40, and 80
- hidden layer size: sampled from the values 128, 256, 512, and 1024

The best results are considered for the final remaining useful life calculation.

Performance Evaluation | The performance is evaluated using the metrics MAE, MSE, and the PHM score, due to their popularity. Table 28 presents the results for the testingS data set. It can be seen that the hyperparameter optimization only slightly improves the results over the baseline model for all four data sets. On the one hand, this is due to the fact that the baseline model already uses optimized parameters, as defined by Wu et al. (2017b). On the other hand, the hyperparameter optimization is limited to the three parameters.

Table 28 | Results of the Vanilla LSTM

	Parameters	Tuned Model			Baseline Model		
		MAE	MSE	PHM Score	MAE	MSE	PHM Score
FD001	Batch size = 5 Learning Rate = 0.717 LSTM = 512	14.83	397.61	828.85	15.15	399.09	1050.38
FD002	Batch size = 5 Learning Rate = 0.9066 LSTM = 512	18.08	619.34	6536.39	17.71	635.26	8195.72
FD003	Batch size = 5 Learning Rate = 0.8064 LSTM = 1024	14.38	394.59	1706.99	16.04	470.46	3073.70
FD004	Batch size = 5 Learning Rate = 0.9939 LSTM = 512	20.45	769.78	6755.43	24.34	875.34	6198.57

The development of artificial neural networks is considerably more complex to implement as compared to the other two approaches. Moreover, more experience in handling and tuning a network is required to achieve good results. Overall, however, high accuracy is reached. To improve the accuracy even further, other network structured could be investigated and tested for their suitability.

5.5.3 Evaluation Discussion

The three instantiations of the algorithm development method demonstrate the applicability of the method with regard to two predefined scenarios. Through the fleet approach selection process, three fleet approaches and one prognostic method are identified, which are implemented subsequently. For each approach, the method further provides a set of steps to be followed. Taking these steps into account, an algorithm is successfully developed. For this purpose, it can be argued that the algorithm development method has been evaluated successfully with regard to the considered data sets and fleet approaches.

It is shown that fleet characteristics are accommodated within all approaches, leading to an increase in performance compared to a simple global model. Out of the three implemented fleet approaches, the similarity-based approaches yields the best overall performance and even out-performes the artificial neural networks. Yet, this result is not surprising as the similarity-based approach has been proven to be highly suitable for this data set and depicts the winning method of the data competition (Ramasso and Saxena 2014, p. 4). In contrast, the local model approach results in the worst prognostic accuracy with regard to all three metrics. But, this can be attributed to the fact that the two resulting homogeneous sub-fleets are targeted by means of a global model. To further improve the quality of the local model approach, fleet approaches for homogeneous fleets should be considered for the two sub-fleets instead.

Despite the successful evaluation of the method, some limitations of the evaluation are pointed out in the following. As the conducted evaluation only depicts three instantiations of the method, it is not possible to draw a conclusion on the general validity of the developed method. In particular, out of the eight fleet approaches, only three have been demonstrated. For this purpose, further instantiations with regard to different data sets and application characteristics should be performed. In addition, the fleet approach selection process could only be evaluated to a limited extent. Based on defined criteria, it is possible to identify fleet approaches. Nevertheless, the suitability of the fleet approaches is not verified yet. Moreover, the assessment of each criterion and fleet approach, respectively prognostic method, can only be evaluated by means of a large number of experiments, which is beyond the scope of this thesis.

5.6 Discussion and Limitations

The processing of fleet data for prognostics, taking into account the machine-specific characteristics, is a recently emerging field of research. The majority of the existing publications are application-specific and therefore not transferable to other cases without adaptation. For this purpose, on the basis of the fleet characterization method, an algorithm development method for fleet prognostics is designed, which provides detailed guidance during the prognostic algorithm implementation with fleet data. The method structures the available knowledge from the literature by means of a generic process, which consists of different steps, their configuration possibilities, as well as decision-making assistance for the guidance of the users during the design of an application-specific solution. Given the fleet prognostic type, the method supports the selection of appropriate fleet approaches and subsequently guides the development of an application-specific fleet prognostic algorithm. The algorithm development method, therefore, satisfies the derived requirements, as presented in Chapter 5.1. It aims to reduce the amount of effort required for the design of a fleet prognostic algorithm by comparing and evaluating the different methods and presenting key implementation steps.

The method contributes to enriching the knowledge base by improving the understanding of fleet prognostics, focusing on the available fleet approaches and their respective characteristics. The method is limited by the current state of fleet prognostics as well as by the selection bias due to the scope of the literature review. Thus, the focus is on the existing identified approaches from the literature and their analysis. However, the method is easily extendable, allowing the integration of further new fleet approaches and methods in the future. Furthermore, the identified steps and implementation details represent only the subset of alternatives covered by the identified literature. Hence, they serve rather as thought-provoking impulses and are not exhaustive. Finally, most of the algorithms are created for one or more C-MAPPS data sets. As a result, the characteristics of this data set are addressed disproportionately often, whereas other fleet characteristics are covered relatively rarely.

The method offers a wide range of recommendations and support for the development of a fleet prognostic algorithm. However, due to the high complexity, the level of recommendation is limited. The user is guided through a variety of processes, with potential alternatives given for the design of each step. With regard to the fleet approach selection process, the fleet approach and general method assessment are based only on a subset of criteria, which have been deemed relevant. In addition, the process of method selection could be further improved by a more structured approach to the exploitation of the qualitative method assessment through, e.g., a criterion-weighted score. For the fleet prognostic implementation process, the identified implementation details are only provided to the user without the specification of key advantages and disadvantages.

Beyond these limitations, two further aspects have to be reflected. This includes the development of a fleet prognostic approach instead of a global model. Even though the global model is not able to capture machine-to-machine variations, it can be a reasonable alternative due to its simplicity. In particular, the option of the global model for identical fleets should be examined and regarded as an equally weighted alternative in the performance evaluation. The second aspect addresses the method selection process. The identified assessments reflect a general evaluation of the method, independent of the method's configuration and fit to the application. The drawbacks and weaknesses of methods can be mitigated or eliminated by specific variants. The decision-support, therefore, mainly addresses the inexperienced users to gain a general understanding of the method's capabilities, whereas the more experienced users should also include their individual expert knowledge into the selection process.

To conclude, the integration of the method with the process reference model is discussed. The method is based on the three processes of data preprocessing, algorithm development, and performance evaluation, as defined by the process reference model on the second level. However, the detailed design with respect to the process elements (third level) differs. The process reference model presents general steps without consideration of specific methods. In contrast, the developed method takes into account the concrete fleet approaches, which leads to the removal (e.g., threshold definition) or insertion of steps (e.g., hyperparameter tuning). However, this does not contradict the content of the process reference model but rather demonstrates the need to adapt the third level for the specific application. Therefore, no adjustments to the process reference model are derived from the developed method.

6 Practical Evaluation with Three Industrial Cases

In the previous three chapters, both the process reference model (cf. Chapter 3) and the fleet prognostic development, which is composed of the characterization method (cf. Chapter 4) and the algorithm development method (cf. Chapter 5), are described. Each of these artifacts is evaluated on its own at the end of the respective chapter. However, in order to evaluate the practical relevance of the overall research results as well as assess the transferability of the results from science to practice, three industrial predictive maintenance projects are realized, in which the obtained results served as a guideline.

For this purpose, the chapter describes and discusses the conclusions drawn from the three evaluation cases. Thus, primarily, the general methodological approach is presented (Chapter 6.1). This is followed by the description and assessment of the three industrial cases (Chapters 6.2, 6.3, and 6.4). Finally, the results are discussed in a summarizing and cross-case perspective (Chapter 6.5).

6.1 Methodological Approach

Design science research demands that research is of high relevance (Hevner et al. 2004, p. 80). In a concluding step, the developed artifacts should therefore be fed back into the environment to be examined and tested in the application domain (Hevner 2007, p. 89). This additional evaluation step allows the artifacts to be evaluated in a more realistic environment that is less artificial than the research environment (Galliers, p. 333). The developed artifacts depict one model and two methods, which define procedures. These artifacts do not have a unique solution, nor can they be labeled right or wrong (Moody and Shanks 2003, p. 624). Practical validity can, therefore, only be shown and achieved through successful industrial applications (Moody and Shanks 2003, p. 624). By instantiating the research artifacts for a particular organizational environment, it is possible to demonstrate the feasibility and applicability (Hevner et al. 2004, p. 79). Therefore, with respect to the main research goal, the practical evaluation should comprise the application of the model and method within organizations to guide their predictive maintenance projects. As a result of these practical evaluations, there may arise a need for further adjustment to improve to the quality of the research artifacts (Hevner 2007, p. 89; Ahlemann and Gastl 2007, p. 93). This practical evaluation, therefore, depicts another iteration of the development and evaluation loop (March and Smith 1995, pp. 258–260; Markus et al. 2002, pp. 193–196), respectively, the relevance cycle (Hevner 2007, pp. 88 f.) of design science research. In addition to the practical evaluation of the three artifacts, the instantiation of the model and method within organizational environments can also yield limitations of the current research and potentials for future research.

For the practical evaluation, the research results are exploited to steer three industrial predictive maintenance projects. Each of the three industrial predictive maintenance projects displays distinct characteristics to allow for a wide range of possible application scenarios. In particular, each of the industrial projects has a different evaluation focus under consideration of the specific application characteristics. The first case is conducted in cooperation with an automotive supplier and focuses on one of their production machines. The second application case targets trailers manufactured by a commercial vehicle manufacturer. Lastly, the third case is carried out with an agricultural machinery manufacturer and investigated one type of machine produced. The first two cases are leveraged to evaluate the developed process reference model. However, for its practical evaluation, it is not mandatory that the entire process reference model is applied in every case; rather, selected parts of the model can be considered (Ahlemann and Gastl 2007, p. 93). This is important as predictive maintenance projects can last several years, from their initial concepts to the final implementation. Consequently, the evaluation focuses on a subset of phases and processes. In particular, the phases of the reference model relevant at the time of the project kick-off are addressed in agreement with the companies. In addition, the second and third case are taken to evaluate the fleet prognostic development method. This involves both the characterization method and the algorithm development method. In contrast to the first case, data from a fleet of machines is available within both of these cases. Table 29 provides a brief summary of the three application cases as well as the evaluation focus with respect to both artifacts.

Table 29 | Evaluation Focus of the Industrial Cases

	Industry	PdM Focus	Process Reference Model	Fleet Prognostics
Case A	Automotive Supplier	Production Machinery	(1) Preparation (2) PHM Design	-
Case B	Commercial Vehicle Manufacturer	Trailers	(1) Preparation (2) System Architecture Design (3) PHM Design	Yes
Case C	Agricultural Machinery Manufacturer	Agricultural Machine	-	Yes

The three predictive maintenance projects are realized in close cooperation with the companies. For this purpose, the role of a consultant is taken with an external perspective to support the customization of the artifacts as well as the realization of the project (Ahlemann and Gastl 2007, p. 93). For each of the industrial projects, the project team is composed of a group of 6-8 students. The general aim comprised the development of a proof-of-concept or a first prototype without operationalization. The realization of the projects is organized into four phases, the project scope definition, the project realization, the project closing, and the project evaluation. For the first phase, the project scope definition, a workshop is conducted with the key stakeholders of the company's project. A workshop depicts a group of employees that come together to work jointly on a topic beyond their daily work routine (Schiersmann and Thiel 2011, p. 96). In particular, the target-finding workshop is selected, which captures the different interests of the stakeholders with regard to the final goal (Schiersmann and Thiel 2011, p. 97). The workshops are designed according to the nine phases of Lipp and Will (2004), namely preliminary contact, introduction, information, target, identification of options and structuring, presentation and discussion of results, evaluation and decision-making, as well as action catalog. Thus, the goal of the workshop is to jointly define the scope and objectives of the project in which the artifacts are applied. In particular, for the process reference model, the relevant phases are identified. After the project scope definition, the project realization phase is carried out. Within this phase, the objectives are worked on in close cooperation with the companies. For this, the applicable artifacts are considered and adapted to the context-specific requirements of the companies. To gather information, a variety of expert interviews and further problem-solving workshops are organized. In addition, an extract of the historical data of the machine(s) under consideration is provided. Through weekly jour fixes, the current results are aligned with the ideas and visions of the stakeholders. After completion of the project, as part of the project closing phase, the results are presented and transferred to the companies, along with detailed documentation of the results. To conclude, in the final phase, the results of the project are evaluated in terms of the applicability of the research artifacts as well as their limitations and need for adjustments.

6.2 Case A: Automotive Supplier

The first application case covers the production line of an automotive supplier. The company attributes great potential for cost savings and increased machine reliability to the implementation of predictive maintenance. For this purpose, a project is initiated that focuses on the implementation of condition monitoring and predictive maintenance for one valuable asset group.

In the following, the project, as well as the derived focus and the application of the research artifacts, are presented, with particular emphasis on the derived procedure model for the company-specific implementation (Chapter 6.2.1). Thereafter, the detailed elaboration of the derived processes and their elements is briefly described (Chapter 6.2.2).

6.2.1 Project Setting and Scope

Case A is developed in cooperation with a company that started working on predictive maintenance two years prior to the collaboration. The company currently performs maintenance by means of pre-defined experience-based intervals. In general, production targets are deemed as most vital; in comparison, maintenance is attributed to low priority. This leads to over-maintenance, which is accompanied by high avoidable costs. Predictive maintenance should, therefore, be introduced as an instrument for more effective maintenance. Predictive maintenance is part of the company's Industrie 4.0 initiative, which aims at improving the scheduling of maintenance activities for cost reduction. For this purpose, a clear vision and initial project deliverables are available. As a result, it was possible to build on various existing results. This includes, in particular, an appropriate design of the system architecture to provide data in an integrated and structured format.

Through the project kick-off workshop, the objectives of the project are identified. In particular, two objectives are jointly defined:

(1) Development of a process for selecting suitable PdM machine candidates

(2) Remaining useful lifetime prediction of a potential PdM machine candidate

The first objective targets a deeper analysis of the available machines and general selection criteria for a systematic assessment of their PdM suitability. In particular, scalability and added value should be focused on. In a second step, a candidate component should be selected, and failure prediction performed through prognostics.

Given the two defined project objectives, the project focuses on the candidate identification process (in the preparation phase) as well as the PHM design phase of the process reference model. As the project targets a proof-of-concept, the deployment phase of the candidate solution is omitted. Furthermore, within the PHM design phase, only the processes of data acquisition/ preparation, algorithm development, and algorithm assessment are considered. This is due to the fact that the development of a predictive algorithm is to be done from a purely data-driven perspective (excluding the process engineering knowledge formulation), and no mitigation of the results is planned (excluding the process solution mitigation). As no fleet data is available for prognostics, the fleet prognostic development method is not taken into account for Case A.

The elaboration of the first objective was grounded in the process of candidate identification. In consensus with the company, the criterion-based assessment approach is chosen and detailed with the input from the experts. The process elements equipment identification, criteria definition, criteria-based assessment, and candidate prioritization are implemented for this purpose. The PHM design phase is consulted for the execution of the second objective. Thus, as stated above, the three processes of data preprocessing, algorithm development, and algorithm evaluation are adopted to design a prognostic algorithm. The implementation of the data acquisition, data understanding, and data preparation process elements enabled the data to be processed for prognostics, which was targeted within the algorithm development process. Within this process, first, anomaly detection is performed, and a threshold is defined. Due to the fact that an artificial neural network is selected as the prognostic method, the historical data is labeled prior. For this purpose, it is required to return to the data preprocessing process. Lastly, within the process algorithm assessment, the performance of the developed prognostic algorithm is evaluated. This resulted in the detailed process model as depicted in Figure 63.

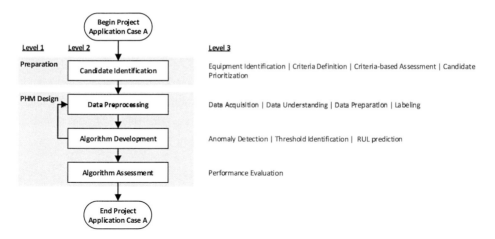

Figure 63 | Detailed Process – Application Case A

6.2.2 Project Description

The design of the processes and process elements are briefly presented in this chapter to provide some more insights into the project results.

Candidate Identification | The first objective is targeted by means of a multi-level process, which depicts a company-specific procedure to classify identified machine candidates for their suitability in the context of PdM. In the first step, available machines and their components are identified, and data availability is checked. Thereafter, exceptions are screened. For this, PdM suitability is directly assigned to candidates with existing customer requirements or PdM use cases as well as machines with process bottlenecks or a high degree of vulnerability. In contrast, machine candidates are excluded with no data availability or small remaining useful life. The remaining subset of machine candidates is classified according to five criteria, namely risk, capacity utilization, degree of standardization, useful life duration, and data availability/ quality. Both criteria and associated weights are formulated in conjunction with the company's experts. A final score of each machine candidate is used to group them into three groups: A (high suitability), B (potential candidates), and C (poor suitability). While the resulting classification presents a snapshot of the current situation, the process should be performed repeatedly to reevaluate adjusted situations in terms of, e.g., new candidates or improved data availability. As a result of this process, the press line and, in particular, the roller conveyor is selected for detailed analysis. This candidate represents a class A machine with long RUL, a high degree of standardization, and sufficient historical data of usage and failure.

Data Acquisition and Preparation | The press line is monitored through 2000 sensors, which are tracked ten times every second. Of these, two sensors are mounted on each of the fifteen roller conveyor actuators to measure both current and speed. In addition, the product number of the transported product is gathered. These data have been further prepared by considering missing values, potential outliers as well as noise. For missing data, linear interpolation is applied due to its robustness and efficiency. Outlier detection is targeted through the upper and lower 5% quantile. As strong deviations can also indicate anomalies, neighboring averages are examined. With this, no clear outlier evidence is identified so that no further changes are implemented. Lastly, the Savitzky-Golay filter is applied for noise reduction. This statistical filter eliminates noise by local polynomial regression and is characterized by its proper treatment of trends.

Algorithm Development | After data was prepared, the health of the machine is assessed. Through visual inspection, trends with sudden drops in the current sensor data are identified. As the peaks matched with performed maintenance activities, the current sensor data is further used. Besides that, the influence of the product is analyzed. To do so, sensor values are grouped based on products, and their averages are evaluated. This revealed only the marginal influence of 4%, which could also be attributed to random effects. Thus, product influence cleaning is not targeted. Additionally, to capture different ranges of speed, the time-series is separated into two (below and above 2.000 RPM). As a physical health indicator, the hourly-aggregated root mean square (RMS) value is chosen due to its simplicity and low computational requirement. The resulting time-series are further examined for different health states through clustering the available data points. For this purpose, k-means is considered with varying values of k. Two clusters (k = 2) are perceived as the best result, which splits data into healthy and faulty behavior. Results are validated by repair data associated with faulty behavior. Given these results, a fault threshold is estimated using a support vector machine. The machine is defined as faulty in case the majority of values are above the threshold within a pre-defined time interval. This interval is set to 8 hours, which corresponds to the duration of one shift. Figure 64 displays the resulting health stage clustering.

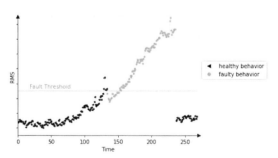

Figure 64 | Health Assessment – Case A

To estimate the remaining useful life, historical data in the faulty state are examined further. Here, artificial neural networks are selected as a prognostic method. To train the network, data labeling is required. For each failure, previous faulty data are labeled in the range [0,1], which depicts the percentage of failure progression. As network architecture, a recurrent neural network with long short-term memory (LSTM) was considered. The network is particularly suitable due to its ability to handle dynamic data. Various network configurations (number of considered past values, number of neurons per layer, number of layers, dropout parameters, and learning rate) are evaluated through cross-validation. The resulting network comprises 6 layers, a learning rate of 0.001, as well as different numbers of neurons per layer and associated dropout rate.

Algorithm Assessment | For network evaluation, the metric R^2 is analyzed. This metric is widely used in regression analysis and can be interpreted as the percentage of explained variance of the independent variable on the dependent variable. The achieved performance is $R^2 = 83\%$. The final RUL can be estimated through extrapolation.

6.3   Case B: Commercial Vehicle Manufacturer

The second application is implemented in cooperation with a manufacturer of commercial vehicles. Besides various trailer variants, the company offers multiple services to its customers. These include, among others, a visualization platform of telematics data and full-service maintenance contracts. To further strengthen existing competencies and remain competitive in the market, the company is exploring additional digital business models. One topic, which emerged in this context is predictive maintenance and related options. For this, the company is particularly interested in the intelligent maintenance of their products, as an added value can be created for the customer, and a large amount of data is already accessible through a telematics system.

In the following, analogously to the first case, the project setting and the case-specific process are presented (Chapter 6.3.1). This is followed by the description of the case (Chapter 6.3.2).

6.3.1 Project Setting and Scope

Case B is conducted with a company, which just started to engage in predictive maintenance and is investigating suitable options, but has not yet initiated a concrete project. Nevertheless, the first steps in this direction are realized. This includes an improved cost report as well as a telematics data visualization portal. Further steps should be geared more strongly to customer benefits and their willingness to pay for the additional service. This is of particular concern as their customers are already struggling with a large number of different products and are therefore only willing to accept additional ones if there is a distinct added value. With regard to maintenance, full-service maintenance contracts for trailers currently cover usage-based preventive and corrective services. Contracts are standardized to a large extent; however, the pricing additionally takes into account, among other factors, past experiences, and mileages. Moreover, data for trailers are already gathered through the telematic and repair system.

The project scope is defined in a joint workshop. In detail, the following project objectives are agreed on:

(1) Elaboration of a PdM Business Model

(2) Development of an Integrated Data Architecture

(3) Prognostic Algorithm Development for Eligible Candidates

In general, the project should cover the exploration of PdM potential as well as a first prototype implementation for promising PdM candidates. For the first objective, initial ideas provided by the company are to be further prepared and devised to provide a better understanding of the future potential. This objective is of particular importance for the company, as initial attempts have met with only limited success in terms of perceived customer benefits. Additionally, with regard to the second and third objectives, the realization of predictive maintenance should also be examined in light of the current data. The second objective, thus, targets the integration of distributed information systems, which are currently siloed. In particular, the repair data is to be linked with the gathered telematics data. The integrated data form the starting point for the third objective. Major components of commercial vehicles are exemplarily examined for their suitability of prognostics and prototypically implemented.

With respect to the three objectives, the realization of Case B targets the preparation, system architecture design, and PHM design phases of the developed process reference model. The first objective, the identification of a business model, can be placed within the process project specification of the preparation phase. For the second objective, a sound system architecture should be established. While the company already maintains different databases, the data is not yet fully integrated. For this purpose, the processes objective and requirements analysis and logical architecture design are to be addressed within this phase. Lastly, the third objective requires the development of prognostic algorithms. For each of the components, the PHM design phase is to be executed with respect to the processes of data preprocessing, algorithm development as well as algorithm assessment. Moreover, as data from a fleet of trailers were available in Case B, the fleet prognostic development method is applicable within the third objective. As the project aimed at a prototypical implementation in order to gain initial insights into the potential of predictive maintenance, the deployment of the results within the company landscape is skipped. This results in the omission of the system implementation process (system architecture phase), the solution mitigation process (PHM design phase) as well as the deployment phase.

The first objective, the development of a business model, is targeted in the preparation phase. Thereafter, a suitable logical system architecture is designed within the system architecture phase, given a short requirement analysis. For the logical architecture design, the process elements data standardization and data integration are focused on. This resulted in an integrated database that is used as the basis for the PHM design phase. In the PHM design phase, the data preprocessing, algorithm development, and evaluation processes are targeted for two pre-selected candidates. Due to the objective of the project, the selection of the two machine candidates is simplified (only expert knowledge and business value) and not accomplished through a comprehensive process as defined in the process reference model. Furthermore, the phase is conducted using the fleet prognostic development method, which is described below in detail. Figure 65 depicts the resulting process for Case B.

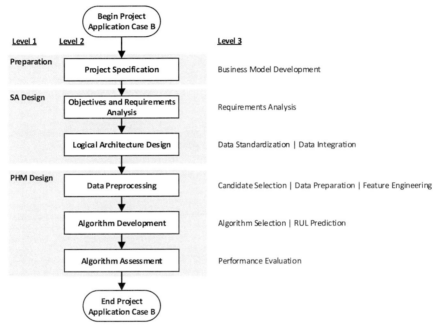

Figure 65 | Detailed Process – Application Case B

As a supplement to the process reference model, for the development of the prognostic algorithm, the fleet prognostic development method is considered. For this purpose, the third objective of Case B was elaborated respecting the developed method. For the implementation, the three identified phases are realized. However, adjustments within the individual phases are essential for the application of the method within this case. As part of the first phase, the fleet is characterized. Given that the data is of high-dimensionality and no condition data is available for neither the tires nor the brakes, only the fleet description step is carried out. As key fleet characteristics, the fleet description revealed different system designs, as well as individual working conditions of the trailer fleet. For this purpose, a heterogeneous fleet is assumed. This is followed by the method selection phase. Here, due to the lack of condition data, a usage-based approach (type-II prognostics) is to be pursued. However, as the fleet prognostic development method targets type-III prognostics, the selection is limited to methods that directly perform RUL mapping. As a result, the artificial neural networks are selected as a fleet method and generalized by also considering further machine learning methods capable of direct RUL mapping. Nevertheless, in order to address the heterogeneity of vehicles due to various working conditions, fleet features are derived in accordance with the method during data preprocessing. These fleet features reflect the vehicle usage pattern. In accordance with the fleet prognostic implementation phase, the steps of data preprocessing, fleet prognostic approach implementation, and performance evaluation are followed. Figure 66 summarizes the resulting fleet prognostic development process.

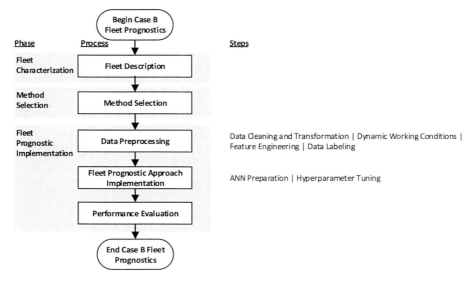

Figure 66 | Resulting Process for Fleet Prognostics – Application Case B

6.3.2 Project Description

In the following, the project is presented in terms of processes and their concrete elaboration in order to provide further details. The processes are highlighted in italic refer to the processes of the process reference model, while the underlined processes correspond to the processes, which are guided by the fleet prognostic development method.

Project Specification | For the realization of the first object, the elaboration of a business model, the roadmap for digital transformation by Schallmo et al. (2017) is considered. As a result, a three-phase procedure is proposed, with each phase generating benefits on its own. Primarily, an intelligent maintenance decision system should be targeted. For this, information obtained from prognostics is considered internally to improve maintenance scheduling, which leads to reduced breakdown and maintenance cost as well as increase planning reliability. This is followed by the second phase, intelligent maintenance contracts, in which the customer is priced based on its usage and corresponding maintenance costs. In the long-term, a self-service customer portal is envisioned, where customers can obtain detailed information on the necessary repair and maintenance services and plan trailer assignments and routes accordingly. This final business model is intended to create a competitive advantage by offering customers an additional benefit and creating new revenue streams for the company.

Objectives and Requirements Analysis | The second objective was targeted by developing an integrated data schema, which covers numerous heterogeneous data sources and transforms them into a unified format by identifying matching criteria. The architecture should be easily re-deployable and be able to update with new data emerging continuously.

Logical Architecture Design | The integration focused on three data sources: (1) the repair system, which processes all repair-related transaction and data, (2) the telematics system, which provides real-time monitoring data as well additional (3) maintenance records for brakes and tires, which provide insights on tire profile depth and brake-pad thickness measured during maintenance checks. Beyond that, additional information is generated, such as the current component installed on the vehicle, component lifetimes, and the route altitude. Furthermore, by accessing an open-source tool, trailer routes are calculated, and the route category (surface condition) is extracted. With respect to these data sources, a logical architecture is designed.

Data Acquisition and Preparation | To develop prototype prognostic implementations, two components, namely tires and brakes, are selected based on expert knowledge.

Fleet Characterization: A large fleet of trailers is available, which is characterized by customized system design and individual usage, and environmental conditions. As the degradation of these components is strongly linked to the operating condition, a heterogeneous fleet is assumed. However, in order to be able to exploit the fleet data within prognostics, a subset of similar trailers is defined based on expert knowledge.

Algorithm Development:

Method Selection: Due to the huge amount of features, the lack of condition data as well as missing domain knowledge, various direct RUL mapping methods are selected for prognostics. Considered methods include linear regression, ridge regression/ classification, gaussian process regression, stochastic gradient descent regression/ classification, k-nearest neighbor, decision tree, random forest, gradient boosting with trees, support vector machines, and multi-layer perceptron for regression.

Data Processing: The resulting data is prepared subsequently. This included the removal of duplications, imputation via forward and backward filling, as well as linear interpolation. As the individual usage of the trailer was named as the main influencing factor for degradation, the working conditions are identified, which is composed of several factors (e.g., driving behavior, road condition, trailer load). This is achieved by clustering various features (including relevant sensors as well as road information) using k-means. Figure 67 depicts the resulting clusters for the average speed as well as load. The cartesian product of all clusters is taken as the operating condition. For tires, an additional working condition was specified based on tire pressure and temperature. To specify the working history, the distance driven in each condition is calculated. The resulting data set is reduced in dimension through PCA. In addition, further features from the vehicle master data were appended and RUL labels calculated. Thereafter, labels are attached in terms of linearly decreasing RUL values. Lastly, k-fold cross-validation is chosen to perform a train-test split of the data.

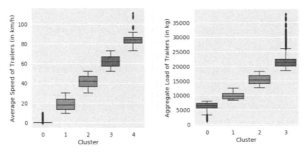

Figure 67 | Feature Clustering – Case B

Algorithm Development (Prognostic Method Development) | The identified methods are implemented, and a preliminary analysis is performed. As a consequence, support vector machine and gaussian process regression are excluded due to extreme computational workload and preliminary average performance on a reduced data set. The remaining methods are tuned via cross-validation.

Algorithm Assessment (Performance Evaluation) | Performance is measured using mean absolute error (MAE). For both tires and bakes, gradient boosting achieved the best results with $mae_{tire} = 117$ and $mae_{brake-pad} = 161$. Similar to the scoring function, an additional analysis of performance over time is conducted. Figure 68 exemplarily depicts the assessment for tires. It can be seen that accuracy increases with failure imminent. The overall rather mediocre performance can be attributed to the lack of exhaustive parameter tuning as well as an insufficient and poor database.

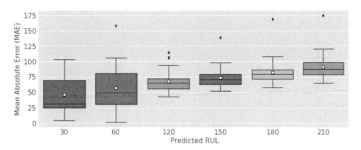

Figure 68 | Performance Assessment of Tires – Case B

6.4 Case C: Agricultural Machinery Manufacturer

The third application case is realized in cooperation with a global manufacturer of agricultural machinery. In this sector, machine reliability for the end customer is very decisive due to a small time window of machine operation throughout the year. Machine failure within this narrow time span generally leads to a major economic loss for their customers. For this reason, the realization of predictive maintenance offers added value for the company, both externally in terms of increased customer satisfaction and improved maintenance planning for the retail network (e.g., in terms of spare parts availability) as well as internally as a result of new business models (e.g., service packages). For this purpose, the company recognizes great potential in further development towards predictive maintenance and has already initiated a predictive maintenance project.

In line with the previous chapters, the project setting and evaluation scope will be presented first (Chapter 6.4.1). Thereafter, the case is described (Chapter 6.4.2).

6.4.1 Project Setting and Scope

The company of case C, similar to case A, already has initial experience in implementing predictive maintenance. The company's predictive maintenance project is initiated one year prior. In the short term, internal cost savings and process optimization are targeted through data analysis, which is followed by improved support for the retailer by means of repair part diagnosis. For this, the retailers will receive further information on the status of the machinery, allowing them to specify the problem in advance and determine the required spare parts and equipment without prior machinery inspection. The ultimate goal is the definition of reliability service packages, which are sold to the end customer. The company already has access to data from the telematics systems of agricultural machines, as well as information on warranty claims and diagnostic trouble codes.

The university project is initiated by means of a kick-off workshop, which is designed to define the target jointly. The workshop resulted in the definition of the scope and target type of machinery. Former is specified by two main objectives:

(1) Characterization of the fleet with emphasis on the vehicle usage

(2) Prediction of a diagnostic trouble code occurrence

The first objective targets a better understanding and characterization of the machinery fleet. In particular, the vehicle usage profiles should be investigated, as it is assumed that usage highly impacts degradation. The second objective targets the prediction of the occurrence of a diagnostic trouble code (DTC). A DTC can be seen as an alert, which indicates a machine fault based on pre-defined thresholds and tolerance intervals. Early detection of a potential DTC supports improved planning in terms of spare parts availability and maintenance schedule, which is particularly important during time intervals of high machine usage. The project aims at developing an initial proof-of-concept, which is further elaborated by the company.

Both objectives can be accomplished by the fleet prognostic development method. While the first objective can be met by the characterization method, the second objective can be realized, taking into account the algorithm development method. Therefore, the focus of Case C is on the practical evaluation of the fleet prognostic development method. The process reference model is not considered here as no further aspects of a predictive maintenance project are addressed beyond the scope of the method.

Within Case C, the three phases of fleet characterization, method selection, and fleet prognostic implementation are realized. Figure 69 shows an overview of the realized phases, processes, and steps. As part of the first objective, the fleet is characterized. For this purpose, the fleet is described with respect to the fleet description, and the fleet type is identified. However, the similarity-based approach, as proposed in Chapter 4.4.2, is not possible owing to the complexity and high-dimensionality of the data set and limited informative value with regard to the individual sensors. In addition, data is only available in an aggregated format, which prevented direct comparison. Thus, from the fleet description, two sources of dissimilarity are derived, the different fault modes as well as the operating condition. As the second objective targeted only one fault mode, the analysis of the fleet type is performed with respect to the identification of different operating condition clusters. For this purpose, the process was adjusted to a feature-based clustering. Even though the resulting Silhouette coefficient was low ($SC = 0.14$), three distinct fleet clusters are identified. With respect to this finding, a heterogeneous fleet type is assigned. This motivated the use of the fleet strategy local models within the algorithm selection phase. Furthermore, a variety of machine learning methods are selected for the implementation of the strategy. This is required due to the lack of explicit degradation trajectories for long-term predictions, which required direct RUL prediction. The results, based on various performance metrics, proved that the local model approach is able to improve the forecast quality over the global model.

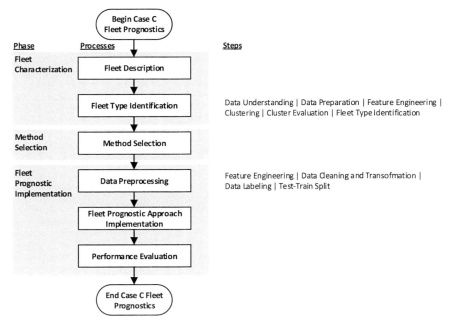

Figure 69 | Fleet Prognostic Development Process Overview – Application Case C

6.4.2 Project Description

This chapter will shortly present the project results. The description is aligned with the phases and processes of the fleet prognostic development method.

Fleet Characterization Method

The first objective targeted the understanding and characterization of the fleet, as well as, in particular, the identification of machine usage patterns. Tractors are chosen as the unit of analysis for reasons of simplicity and data availability. The transfer of results to more complex harvesters is envisioned within a later phase.

Fleet Description: The targeted fleet comprises 39 vehicles of one specific type of tractor. Thus, the system design can be described as identical. However, vehicles differ greatly with regard to their operating condition. In addition, multiple fault modes are available.

Data Understanding: The telematics system collects data in snapshots every few seconds during machine usage. The data resides prepared and harmonized in a Hadoop cluster. Within the project, an extract of 11 months of the database is provided, with ~35 million observations. Each observation included 46 features, which can be grouped into time-related (e.g., measurement time), categorical (e.g., vehicleID, country), and numerical (e.g., temperature, speed) features. This data set is investigated in more detail using data profiling. Results showed various descriptive statistics for each feature, including the percentage of missing values, min, max, and mean. Besides a deeper understanding of the data, this analysis led to the identification of outliers as well as features with a high amount of missing values. Moreover, related diagnostic trouble codes are provided as well as a short description. By focusing only on critical events, ~25 thousand errors or warnings are identified.

Data Preparation: Thereafter, data preparation in terms of cleaning and reduction is performed to transform the data into a suitable format. At first, outliers, which are identified during profiling, are checked against expert knowledge and removed in case of erroneous entries. In addition, features with a high amount of missing values (> 90%) are deleted. In the course of missing data analysis, it became apparent that the availability of measures depends on the transmission system.

Feature Engineering: For this purpose, various features are defined. Initially, a sub-set of eleven relevant sensors are identified based on expert knowledge and variability among machinery (measured through boxplots). To aggregate these sensors both by day and year, several features are defined, which are classified into descriptive statistic features (mean, median, standard deviation, min, max, and quantile), operating times (sum), gradient features, local extreme point features, as well as start/ stop features. All features are standardized. Due to the huge amount of resulting features (120), principal component analysis is applied to reduce the dimensions further.

Clustering: For clustering, the first five principal components are considered as they explain more than 95% of the variance among the vehicles. Various clustering methods are implemented (k-means, MeanShift, DBSCAN, agglomerative hierarchical clustering).

Cluster Evaluation: For this purpose, three internal cluster evaluation metrics are analyzed (Silhouette coefficient, Calinski-Harabasz index, and Davies-Bouldin index), and the resulting best parametrization and method is chosen. This process resulted in three clusters defined by k-means. In comparison to the other methods, k-means is less sensitive to outliers and therefore identified three reasonable clusters with regard to the targeted objective. Furthermore, both daily and yearly data aggregations are analyzed. Due to a higher degree of separability, yearly aggregation is opted and further analyzed.

Fleet Type Identification: Subsequently, the identified clusters are interpreted with regard to the target. Among others, box plots of the most influential features are examined (cf. Figure 70). From this analysis, the following clusters can be derived with respect to the features engine speed and load and mean engine fuel rate: cluster 0 (low usage), cluster 1 (high usage), cluster 2 (moderate usage). Given substantial differences between the clusters, the heterogeneous fleet prognostic type is assigned.

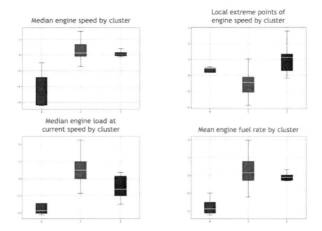

Figure 70 | Distribution of Most Influential Features among Clusters – Case C

The second objective considers the prediction of one crucial DTC. For this purpose, an appropriate DTC candidate is first selected. Thus, a subset of critical and valuable DTCs is defined by the company's expert. This subset is subsequently analyzed with regard to the frequency of occurrence within the provided data. Based on these criteria (criticality, business value, and occurrence), the most promising DTC is chosen.

Algorithm Development Method

Algorithm Selection: With respect to the identified heterogeneous fleet type, the local model approach is selected. Furthermore, for the implementation of each local model, various machine learning methods are chosen due to the lack of a strong health indicator and thus the need for a direct RUL mapping approach. Benchmark methods are XGBoost, SVM, Linear regression, Kneighbor Regressor, GPR, Decision Tree, Random Forest, and Multi-Layer Perceptron.

Data Preprocessing: For feature engineering, in total, 530 features are considered based on gradients of aggregated features, the number of relevant DTC observations within a pre-defined time interval as well as hourly (8 hours in the past) and daily (last seven days) aggregated data of the main features (as used in machine usage profile clustering). This is followed by test design preparation through data labeling (remaining time until occurrence) and train-test split (cross-validation). For the former, time-series were segmented at the time of the DTC occurrence and were labeled for each segment backward, starting with the time of occurrence.

Fleet Prognostic Approach Implementation: Given that DTC predominantly occurred within vehicles assigned to cluster 1, two models are implemented: (1) one only for cluster 1 as well as (2) a global model. The models for the other two machine usage clusters are neglected as the amount of data was insufficient. Nevertheless, the DTC occurrence for vehicles in these clusters could be predicted using the global model. For each method, hyperparameter tuning is iteratively performed through a random search. As XGBoost consistently performed best, both models are implemented using XGBoost.

Performance Evaluation: The performance of the two models is assessed based on three commonly used performance measures in prognostics: (1) the mean average error, (2) the rate of acceptable predictions as well as (3) the timeliness measure. The local model for cluster 1 outperformed the global model in all measures. As an example, the mae is even almost halved as a result (global model: 284h, local model: 149h).

6.5 Practical Evaluation and Discussion

To conclude the practical evaluation, the applicability of the artifacts within the three cases is reviewed hereafter in order to draw conclusions on the utility in practice as well as on potentials for improvement. This is accomplished by first presenting the results and conclusions for the artifacts separately, followed by an overall discussion of the practical evaluation.

The first artifact, the *process reference model*, is evaluated within Case A and Case B. The two cases each cover a part of the process reference model. In particular, the preparation, system architecture design, and PHM design phases are considered, while the candidate deployment and project closure phases are excluded. The latter is due to the consulting perspective taken, leaving the deployment in the corporate landscape to the responsibility of the companies. For the other phases, the processes considered are tailored to the current project status of the company with regard to the defined project objectives. An application-specific process is derived from the scope of each project. Regarding the addressed parts from the process reference model, some general conclusions can be drawn for its practical evaluation. It has been demonstrated that the process reference model is suitable to provide guidance in designing an application-specific predictive maintenance implementation process. The projects are successfully carried out respecting the process flow defined by the derived application-specific model. The model has outlined the process in great detail and provided step-by-step recommendations. In addition, due to the very different application characteristics, the generality of the model can be emphasized. However, this high generality is accompanied by an increased effort for adaptation. In particular, it became apparent that predictive maintenance projects often only require a small subset of processes and process elements. However, due to the reference nature of the model, this does not depict a limitation. Nevertheless, further support for deriving an application-specific model could increase the overall usability. Besides this general finding, two deviations from the model are identified. On the one side, this depicts the cyclic relationship between the preparation process and the algorithm development process, which is observed in Case A. To better accommodate this iterative loop, a return arrow should be added at the process level between these two processes (as highlighted in the application-specific model in Figure 63). On the other hand, for

Case B, the process element business model development is missing in the process reference model. Despite the fact that this step is crucial for the realization of business value from the implementation of predictive maintenance, it is not explicitly represented. However, this is caused by the fact that the process reference model primarily covers predictive maintenance for production machines. Within the setting, the emphasis lies on improving maintenance and machine reliability rather than on generating revenue streams. The elaboration of a business model is therefore dispensable. For this reason, given the scope of the process reference model, it is refrained from extending it by the process element business model development. In conclusion, it can be argued that both cases represent successful validations of the process reference model despite the discussed minor limitations and derivations.

The *characterization method for fleet prognostics* is applicable within the Case B and C. Both cases covered the fleet description, whereas the fleet prognostic type identification process is only considered within Case C. Due to the lack of condition data in Case B, the fleet prognostic type could only be conceptually derived from the fleet description. For both cases, the method is used to provide recommendations for systematic analysis and understanding of the fleet, with the purpose of defining a fleet prognostic type. The fleet description is particularly helpful in this regard, as it provided a quick and effortless way to better understand the fleet and its key characteristics. The fleet prognostic type identification process has to be modified in order to be applicable for Case C. This is required due to the high complexity and dimensionality of the data set as well as the availability of aggregated values, which prevented direct comparison of time-series. As a result, the approach is adapted using feature-based clustering. From this, a fleet type is identified along with more homogeneous sub-fleets. Given this limited applicability of the time-series clustering approach, a need for a more generic clustering approach can be derived, which can better cope with the defined data characteristics. The realized feature-based clustering approach provides an option to overcome this limitation. However, this approach leads to information losses. It should, therefore, only be used in case the implementation of the developed time-series clustering approach is not applicable. Given these results, for the characterization method, however, no definite conclusion can be drawn in terms of practical validity. The fleet description has been successfully applied, while the identification process required adjustments. Nonetheless, for both cases, it is possible to gain information on the fleet as well as identify a fleet prognostic type. In this regard, the method generates an added value for the realization of practical applications and can thus be evaluated positively.

The third artifact, the *algorithm development method for fleet prognostics*, is used within Case B and Case C. The method is leveraged to guide both algorithm selection and algorithm development. First, a suitable strategy and method are selected, taking into account the elaborated method assessment criteria. Second, a fleet prognostic algorithm is developed following the respective processes. The method is able to provide detailed recommendations within all steps of the algorithm development. Within Case C, it is even shown that the fleet prognostic method is able to outperform the global model. Nevertheless, one limitation of the method is observed within the practical evaluation. The limitation refers to the fact that the data characteristics of the two cases are yet not comprehensively covered by the method. For Case B, this is due to the lack of condition data. Here, only a type-II prognostic approach is applicable, which required a direct RUL mapping approach. Likewise, in Case C, no strong degradation trajectory is identified in the data, which also precluded the applicability of a degradation-based approach. However, this decision criterion is currently not included within the method selection process. In order to facilitate the selection and provide additional decision support, the method selection process should be extended with the criteria RUL mapping approach. Furthermore, it is observed that methods for direct RUL predictions are underrepresented within the algorithm development method. The practical evaluation, however, provides an indication that direct RUL prediction could be of high relevance for practice, whereas research primarily focuses on degradation-based prediction. From this, a possible need for research in the direction of direct RUL mapping approaches can be derived, which should subsequently be extended within the algorithm development method. In light of these findings, it can be claimed that Case B and Case C represent successful validation of the method. Moreover, the proposed extensions would increase the quality of the method. The identified improvement potentials for the artifacts, as well as the status of the practical evaluation, are summarized in Table 30.

Table 30 | Key Findings from the Practical Evaluation

	Process Reference Model	Characterization Method	Algorithm Development Method
Case A	Adapted Process Flow	- Not applicable-	- Not applicable-
Case B	-		• RUL mapping criterion
Case C	- Not applicable -	Feature-based Clustering Approach	• Focus on RUL mapping approaches
Evaluation Status	Twice validated	Utility demonstrated	Twice validated

In the practical evaluation of the three artifacts, practical validity is confirmed for the process reference model and the algorithm development method in two cases. In addition, the utility of the characterization method is demonstrated twice. This can therefore be seen as a first successful practical evaluation. The evaluation is, however, limited in the number of cases, as well as the scope of each case. In particular, the evaluation of the process reference model targets only some parts of the overall process. In addition, as the projects are conducted by the researchers, the handling and application of the artifacts by practitioners are not studied. Thus, despite the promising results, further cases, as well as action research, should be targeted to conclusively validate the three artifacts.

7   Conclusion

The implementation of predictive maintenance as a proactive maintenance approach has become increasingly popular in the era of digitalization and the fourth industrial revolution. While research on predictive maintenance and prognostics and health management has been progressing for more than a decade, a gap between research and industrial applications is identified. Most companies are currently addressing this topic and are often only engaged in developing initial solutions. To foster the effective implementation of predictive maintenance, supportive guidance for the realization of a predictive maintenance project is needed that can meet the identified challenges. This support can be provided by a process-centric view of the implementation of predictive maintenance, which can be exploited to plan and execute predictive maintenance projects.

This chapter concludes the thesis. For this purpose, in Chapter 7.1, the key research findings are shortly summarized, juxtaposed with the derived research goals, and their interrelations are presented. Thereafter, in Chapter 7.2, the overall limitations are outlined. Finally, an outlook and avenues for future research are outlined in Chapter 7.3.

7.1   Summary

Implementation of predictive maintenance in practice is often complicated by several hurdles. Two prevailing challenges are the missing systematic guidance as well as the scarcity of real machine data. Existing research in predictive maintenance, however, is, for the most part, explorative and application-specific. Although a large body of knowledge already exists, it is not sufficiently tailored towards the practical challenges. For this purpose, the major research goal of this thesis is to fill this gap by investigating both challenges and provide processes to guide the effective implementation of predictive maintenance and fleet prognostics through a process-centric view. Given this main research goal, three interrelated research questions and goals are formulated, which are answered by means of three artifacts (cf. Chapter 1.2).

In order to attain the main research goal, first, the realization of a predictive maintenance project needs to be systematized and structured. This is targeted through the development of a process reference model for predictive maintenance (*Research Goal 1*). The developed process reference model provides comprehensive assistance and structured guidance for the realization of predictive maintenance for both practitioners and researchers through the provision of phases, processes, and process elements. The model is developed following a rigorous research methodology and consequently draws on the theoretical body of knowledge in predictive maintenance as well as the four related fields of data analytics, business process improvement, systems engineering, and project management (cf. Chapter 3.3). From this, a process model is deduced, which is structured in a hierarchical manner in three levels of detail (cf. Chapter 3.4). On the highest level, a phase model is developed, which provides five generic phases. These phases are detailed on the second level as concrete processes for each defined phase. On the lowest level, specific steps and tasks executed for predictive maintenance implementation are outlined. The applicability of the model, as well as the compliance with multiple proposed requirements, is demonstrated as part of an analytical and empirical evaluation with industrial insights from eleven expert interviews (cf. Chapter 3.5). Due to its reference character, the process reference model for predictive maintenance describes a generic and application-independent view of the realization of a predictive maintenance project. It serves as a basis for the creation of an application-specific model, considering the relevant application characteristics and requirements. The process reference model thus paves the way for the systematic implementation of predictive maintenance projects both in practice and in research.

The process reference model addresses the first practical challenge regarding the lack of required systematic guidance for predictive maintenance implementation by providing an exhaustive overview of the overall process of predictive maintenance projects. By this, it accomplishes the first research goal. The process reference model, however, does not offer any further support for the concrete design of the processes. Therefore, it must be supplemented to overcome the second practical challenge of data scarcity. For this challenge, the research field of fleet prognostics is examined. To this end, recommendations should be given for the development of a fleet prognostic algorithm during the key phase of the process reference model, the PHM design phase. This requires characterizing fleets and their use for prognostics, as well as guiding the development of a fleet prognostic algorithm considering the fleet-specific requirements.

For answering the second and third research goal, the development of a method is targeted for algorithm development for fleet prognostics. This artifact is composed of two sub-artifacts, the characterization and the algorithm development method for fleet prognostics. The characterization method (*Research Goal 2*) should facilitate the description and delimitation of fleet data, from which prognostic requirements are derived. These requirements are addressed within the algorithm development method (*Research Goal 3*). Here, a systematic procedure is aimed at easing the design of an algorithm considering fleet data.

In light of the main research objective, the second research objective targets the development of a characterization process to support the identification of fleet characteristics as well as the derivation of fleet prognostic requirements. Based on a comprehensive review of the knowledge base, the fleet is first conceptualized for its usage within prognostics by identifying key characteristics (cf. Chapter 4.3). Building on these findings, a characterization method for fleet prognostics is derived (cf. Chapter 4.4). The developed method comprises two steps. The first process step addresses a conceptual fleet description. Based on four dimensions, an improved general understanding of the fleet is envisioned. Subsequently, a fleet prognostic type can be assigned to the fleet. For this purpose, an existing categorization is extended with a data-driven perspective along with a process for data-driven fleet prognostic typification. The method is evaluated by means of thoroughly explored data sets (cf. Chapter 4.5). Despite the generality of the method, concrete recommendations are highlighted. However, its realization is application-specific, requiring possible adaptations. For this reason, a full factorial analysis is implemented to assess the impact of different design options.

The developed characterization method provides means to analyze fleet data for prognostics both from a conceptual as well as a data-driven perspective. Thus, it allows to delineate fleets other and determine their main characteristics, which can be used as requirements for the development of a fleet prognostic algorithm. The method developed as part of the second research goal lays the foundation for better structuring the research field of fleet prognostics to steer future research toward key fleet characteristics. By this, it encourages application-independent research, which can be transferred more easily to applications with similar characteristics. Moreover, the method supports the realization of industrial fleet prognostic applications in determining the feasibility as well as deriving fleet prognostic requirements through a comprehensive process. As a final outcome, the characterization method enables the identification of crucial features that need to be considered in fleet prognostics. Therefore, it can be concluded that it fulfills the second research goal.

As part of the third research goal, the development of an algorithm for fleet prognostics is supported, which is leveraged by the elaborated fleet prognostic requirements. It, therefore, builds on the characterization method. Respecting the main research goal, an algorithm development process is developed, which provides guidance both during the selection of a fleet prognostic method and during its implementation. For this purpose, the existing knowledge base is thoroughly examined, and fleet prognostic approaches are identified (cf. Chapter 5.3). In view of these approaches, the algorithm development method for fleet prognostics is elaborated (cf. Chapter 5.4). It is based on two key processes. As part of the fleet approach selection process, an assessment of fleet approaches as well as general methods is provided. Besides the fleet prognostic type, the assessment comprises additional relevant criteria. Thereafter, the fleet prognostic implementation process outlines the required steps and possible design options with respect to data preparation, fleet prognostic approach implementation, and performance evaluation. The application of the method is successfully demonstrated within two scenarios (cf. Chapter 5.5). In particular, it has been observed that incorporating fleet requirements into the development of fleet prognostic algorithms results in an improvement in accuracy compared to a global model.

The devised algorithm development method facilitates the implementation of a fleet prognostic algorithm by reducing its implementation effort. A general process is formulated and the conduction of the steps supported, allowing application-specific adjustments. As a result, it can be concluded that the third research objective of this thesis is achieved. Along with the characterization method, the algorithm development method contributes to the advancement of the research field of fleet prognostics. The focus is shifted from explorative application-specific research towards more generic, application-independent research, which is crucial to the further development of the field. For practitioners, the methods provide detailed and effective guidance for the development of a fleet prognostic algorithm. In addition, it tackles the challenge of the no-free-lunch-theorem by providing a multi-criteria approach to algorithm selection. Thus, practitioners can benefit from both a generic and application-independent view of fleet prognostics as well as detailed recommendations for implementing an application-specific algorithm.

As a final step, in view of the development and evaluation cycle of design science research, a practical evaluation is realized (cf. Chapter 6). For this purpose, the developed artifacts are instantiated in a more realistic environment by means of three application cases. With respect to the three cases, the applicability and value of the artifacts in practice are demonstrated. In addition, potential improvements are identified to render future research more relevant to practical applications.

In summary, the achieved results of this thesis fulfill the three derived sub-goals and, therefore, the main research goal. By means of a wide range of processes at varying levels of detail, the implementation of predictive maintenance and fleet prognostics is supported. Figure 71 provides an overview of the developed artifacts and their interrelations.

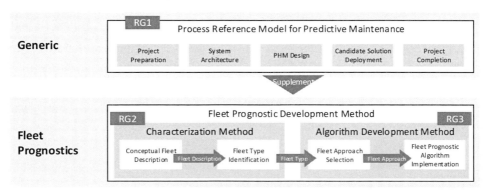

Figure 71 | Overview of Research Results

7.2   Limitations

The thesis bridges research and practice by providing application-independent, generic guidance for the realization of predictive maintenance projects. Despite the described contributions, a few limitations can be derived.

The *process reference model* is developed to provide assistance and structured guidance in designing a predictive maintenance implementation process. It is built on a large body of knowledge. However, given the non-standard terminology in the field and the resulting complexity of identifying the knowledge base, it can be stated that the literature review remained at the level of representative coverage and is thus not fully comprehensive. Furthermore, due to the high diversity of characteristics of predictive maintenance implementation, a trade-off is required between the general validity and usefulness of the model. For this purpose, the first two levels (phase and process level) of the process reference model target a high level of abstraction and thus remain to a large extent generic and application-independent; the third level serves more as a reference point which requires considerable effort during model customization. However, supporting the derivation of an application-specific model taking into account the characteristics of the application domain is beyond the scope of this thesis. In particular, further guidance during the model customization, as well as for the detailed implementation of the processes and process elements, could improve the applicability and useability of the process reference model for practitioners. In addition, the process reference model is limited by its scope. In particular, it focuses on the process flow and highlights main documents within key phases of predictive maintenance implementation, neglecting further perspectives on the predictive maintenance implementation process. Lastly, the evaluation depicts another limitation. Even though an analytical and an empirical evaluation is performed, no claim to evaluation completeness can be asserted. Further forms of evaluations could strengthen the proposition for validity and suitability.

The *characterization method for fleet prognostics* is developed to identify specific fleet requirements which should be addressed in prognostics. For this purpose, a general fleet description and a fleet type identification process are derived. The dimensions and categories of the general fleet descriptions are limited to those reported in the identified literature. The fleet description is therefore not exhaustive with respect to the possible dimensions and criteria of fleets. This is, on the one hand, attributed to the literature search scope and, on the other hand, to the immaturity of the research field. Furthermore, the main limitation of the fleet prognostic type identification process is the strong focus on the similarity of the degradation behavior of machines within a fleet. The process assumes the existence of structurally identical data for each machine in the fleet (e.g., in terms of available sensors, acquisition time), as well as data exhibiting a clear degradation trend (either through a physical or virtual health indicator). This assumption, however, does not always hold in practical applications. In particular, in big data environments, fleet data might exhibit velocity and variability, limiting the applicability of the method in these cases.

This limitation becomes apparent within the two application cases, which prevented the application of the process. Furthermore, the method leverages time-series clustering for fleet type identification. This is beneficial in cases with no prior knowledge of the fleet. However, whenever additional information on the fleet and its composition is available, this data could also be integrated within the fleet data analysis. As a final limitation, the clustering of time-series induces a high degree of complexity. In order to ease the identification of the fleet type, it might often be sufficient to visualize the data in a suitable format, e.g., by time-series plotting.

The *algorithm development method* for fleet prognostics aims to guide the development of a fleet prognostic algorithm taking into account the derived fleet prognostic types. To do so, the method comprises two phases: fleet approach selection and fleet prognostic implementation. The former presents a set of criteria against which the fleet approaches are assessed. Even though the various assessment criteria provide a multi-criteria decision space for method selection, only general method capabilities have been revealed. This is mainly caused by the fact that the capabilities of a method strongly depend on the actual implementation. In addition, variants of methods have been neglected, which could also exhibit divergent characteristics. Furthermore, the criteria, as well as the assessment, are rigorously drawn from the literature. However, due to the huge amount of method and data characteristics, it is beyond the scope of this thesis to validate the different assessments experimentally. The limitation of the fleet prognostic implementation process can be mainly attributed to the literature review process. While the search is exhaustive with regard to the defined keywords, it is not comprehensive in terms of the overall research field. In particular, approaches not using the term fleet or a multitude of further synonyms have not been identified. This selection bias results in a limited inclusion of so-called black-box methods. These methods learn a mapping between the input and output data without the need to explicitly define the relationship. As a result, the majority of identified approaches apply degradation-based mapping, whereas direct RUL mapping was not focused frequently. While this is appropriate in cases where a degradation trajectory can be extracted from the data, a direct RUL mapping must be performed in the remaining cases. This limitation is apparent within the two application cases, which required significant adaptations to direct RUL mapping.

The *experimental evaluation* of the characterization method and the algorithm development method is conducted using four turbofan engine degradation data sets. For each of the data sets, the artifacts are instantiated. However, despite exhibiting different fleet characteristics, the data sets share many similarities. Within the evaluation, it is demonstrated that fleet characteristics could be adequately addressed by the artifacts. Although the results confirm the usefulness and functionality of the artifacts for these data sets, application with respect to other data sets with different characteristics could improve the evaluation and thus ensure the generalizability of the achieved results to a wider range of fleet prognostic applications.

The *practical evaluation* is limited in two ways: the number of cases and the pursued scope. Even though three industrial cases have been selected, which cover a broad spectrum of characteristics, they represent only a small subset of possible application scenarios. Moreover, the three cases are restricted to a small part of the complete project due to the time length of over several years. Thus, it can be argued that the practical evaluation is not exhaustive and cannot be characterized as a comprehensive and complete evaluation. For this purpose, the results are rather indicative and can only be generalized to a certain extent. The second major limitation of the practice evaluation relates to the fact that the implementation and thus the application of the artifacts is not performed by the practitioners. Instead, the projects are realized by researchers who adopt the role of external consultants. Thus, while it is possible to evaluate the implementation of the research artifacts in practice, it is not possible to study the practitioners' handling and application of the artifacts. One way to counteract this limitation is action research.

7.3 Outlook

The results of this thesis contribute to improving the accessibility of predictive maintenance for industrial applications by supporting its implementation. This thesis represents the first step in closing the current gap between research and practice. Thus, several starting points for future work can be derived from both the limitations of this work and the general observations and insights from the research field to advance the implementation of predictive maintenance in the industry.

Concerning the *process reference model*, additional guidance in adapting the model to the application-specific characteristics and in implementing the process elements would increase the value and usefulness of the model. For this purpose, a mapping between application characteristics and relevance of processes and process elements could be targeted, in addition to a detailed description (e.g., task, input/ output, reference to related publications and standards) of each process element. Furthermore, as the developed process reference model primarily emphasizes the general process flow, further process modeling perspectives would create additional benefits, including, among others, the organizational unit (responsibility) as well as the support tools. These extensions would further ease the customization of the process reference model as well as extend its scope. As a result, the implementation of predictive maintenance would be further systematized, and thus more assistance is provided for practitioners in designing their predictive maintenance implementation projects.

To increase the general applicability of the *characterization method for fleet prognostics,* it should be extended considering the identified limitations. In particular, the fleet prognostic type identification process could be augmented to accommodate fleet applications in which time-series clustering is not applicable. Toward this end, a feature-based clustering approach could be designed, which provides another means to identify fleet prognostic types. In addition, the handling of structural data differences within the fleet could be investigated, which is omitted in this thesis. Lastly, the method is developed with a focus on prognostics. However, given the relevance of fleet data for fault detection and diagnostics, it could be investigated how far results are transferable and, from this, derive the corresponding required adjustments. Each of these extensions would broaden the scope and thus the applicability of the developed method.

The research conducted within this thesis has contributed significantly to the advancement of the field of fleet prognostics. Nevertheless, there is still a great potential for further research in this area. This can furthermore be illustrated by the *algorithm development method for fleet prognostics*. Despite the large amount of literature that is identified, further development of methods and algorithms that explicitly account for the highlighted fleet characteristics is encouraged. In particular, it is recommended to place emphasis on methods that are able to perform direct RUL mapping. Although this has been highlighted as very relevant in practice, it has not yet been exhaustively addressed in academic research. In addition, an experimental investigation of the general method properties would be beneficial to further support and fortify the method selection. This thesis provides the groundwork for this investigation through a conceptual perspective. However, experimental verification of the various assessments remains an open aspect.

Design science research requires several iterations of the *development and evaluation loop*. Within this thesis, the three artifacts have been evaluated individually and jointly as part of the practical evaluation. To further evolve the artifacts, more iterations should be implemented to promote continuous improvement. For the process reference model, this implies the execution of further evaluation types; likewise, for the fleet prognostic development method, additional experimental evaluations should be conducted using other data sets. Beyond the individual evaluations, it would be valuable to further investigate and explore the applicability of the artifacts in practice by means of action research.

The research has been conducted with respect to two prevailing challenges, which resulted in the developed process reference model as well as the fleet prognostic development method. While the former is generally applicable, the latter addresses one specific aspect within predictive maintenance implementation and therefore supplements the process reference model. However, practical applications currently face many more challenges. As part of the conducted expert interviews and the implementation of the three evaluation cases, three further practical challenges have been identified, namely, (1) definition of a business model/ case, (2) conduct of a cost-benefit analysis, and (3) change management due to reluctance and reservation to the topic. These challenges present research opportunities for future work, as these areas are not thoroughly explored yet. For this purpose, the process reference model could be supplemented with additional methods or models which target one or multiple of these challenges more in-depth.

Finally, the research field would benefit significantly from an increased focus on standardization. This includes, first, the highly inconsistent use of terminology, which has caused considerable difficulty in identifying the appropriate literature. A clear delimitation of the key terms and the adherence to this terminology would provide new researchers and practitioners with an easier entry point into the research field. Secondly, given that research is currently, for the most part, explorative and application-specific, comparability between the different results should be established. This could be done by the definition of standard research guidelines, which specify key attributes that should be described as well as common processes and metrics that should be followed. Apart from facilitating the comparison of research results, this standard procedure could also foster the identification of related contributions.

To conclude, the artifacts developed in this thesis meet the research questions raised, and each artifact contributes to both the research community and practitioners. In addition, opportunities for further research are identified that would both generate additional benefits for practice and further advance the research field.

References

Abanda, Amaia; Mori, Usue; Lozano, Jose A. (2019): A Review on Distance Based Time Series Classification. In *Data Mining and Knowledge Discovery* 33 (2), pp. 378–412.

Adesola, Sola; Baines, Tim (2005): Developing and Evaluating a Methodology for Business Process Improvement. In *Business Process Management Journal* 11 (1), pp. 37–46.

Adhikari, Ratnadip; Agrawal, R. K. (2013): An Introductory Study on Time Series Modeling and Forecasting: LAP LAMBERT Academic Publishing.

Adolfsson, Andreas; Ackerman, Margareta; Brownstein, Naomi C. (2019): To Cluster, or not to Cluster: An Analysis of Clusterability Methods. In *Pattern Recognition* 88, pp. 13–26.

ADS 79D-HDBK, 2013: Aeronautical Design Standard Handbook.

Agarwal, Vivek; Lybeck, Nancy J.; Pham, Binh T.; Bickford, Randall; Rusaw, Richard (2015): Implementation of Remaining Useful Lifetime Transformer Models in the Fleet-Wide Prognostic and Health Management Suite. In : Proceedings of International Conference on Nuclear Plant Instrumentation, Control and Human-Machine Interface Technologies, vol. 9. Charlotte, United States, pp. 23–26.

Agarwal, Vivek; Lybeck, Nancy J.; Pham, Binh T.; Rusaw, Richard; Bickford, Randall (2012): Prognostic and Health Management of Active Assets in Nuclear Power Plants.

Aghabozorgi, Saeed; Seyed Shirkhorshidi, Ali; Ying Wah, Teh (2015): Time-Series Clustering – A Decade Review. In *Information Systems* 53, pp. 16–38.

Ahlemann, Frederik; Gastl, Heike (2007): Process Model for an Empirically Grounded Reference Model Construction. In Peter Fettke, Peter Loos (Eds.): Reference Modeling for Business Systems Analysis: Idea Group Publishing, pp. 77–97.

Aizpurua, Jose I.; Catterson, Victoria M. (2015): Towards a Methodology for Design of Prognostic Systems. In : Proceedings of Annual Conference of the Prognostics and Health Management Society, vol. 7. Coronado, USA, pp. 504–517.

Aizpurua, Jose I.; Catterson, Victoria M. (2016): ADEPS: A Methodology for Designing Prognostic Applications. In : Proceedings of European Conference of the Prognostics and Health Management Society, vol. 3. Bilbao, Spain.

Al-Dahidi, Sameer; Di Maio, Francesco; Baraldi, Piero; Zio, Enrico (2016): Remaining Useful Life Estimation in Heterogeneous Fleets Working Under Variable Operating Conditions. In *Reliability Engineering & System Safety* 156, pp. 109–124.

Al-Dahidi, Sameer; Di Maio, Francesco; Baraldi, Piero; Zio, Enrico (2017): A Locally Adaptive Ensemble Approach for Data-Driven Prognostics of Heterogeneous Fleets. In *Proceedings of the Institution of Mechanical Engineers, Part O: Journal of Risk and Reliability* 231 (3).

Alpaydin, Ethem (2019): Maschinelles Lernen. 2. Auflage. Berlin, Germany, Boston, United States: De Gruyter Oldenbourg.

Anger, Christoph (2018): Hidden Semi-Markov Models for Predictive Maintenance of Rotating Elements. TU Darmstadt, Darmstadt, Germany.

Antunes, Cláudia; Oliveira, Arlindo (2001): Temporal Data Mining: An Overview. In *KDD Workshop on Temporal Data Mining*, pp. 1–13.

Atamuradov, Vepa; Medjaher, Kamal; Dersin, Pierre; Lamoureux, Benjamin; Zerhouni, Noureddine (2017): Prognostics and Health Management for Maintenance Practitioners-Review, Implementation and Tools Evaluation. In *International Journal of Prognostics and Health Management* 8 (060), pp. 1–31.

Awad, Mariette; Khanna, Rahul (2015): Efficient Learning Machines. Theories, Concepts, and Applications for Engineers and System Designers. New York, United States: Apress Open.

Babu, Giduthuri S.; Zhao, Peilin; Li, Xiao-Li (2016): Deep Convolutional Neural Network Based Regression Approach for Estimation of Remaining Useful Life. In : Proceedings of Database Systems for Advanced Applications. Dallas, United States: Springer, pp. 214–228.

Bacchetti, Andrea; Saccani, Nicola (2012): Spare Parts Classification and Demand Forecasting for Stock Control: Investigating the Gap Between Research and Practice. In *Omega* 40 (6), pp. 722–737.

Bachmair, Dominik (2018): Der Weg zu predictive Maintenance in der industriellen Serienfertigung. Ein Vorgehensmodell zur Einführung von predictive Maintenance auf Basis von Business Intelligence. Beau Bassin, Mauritius: AV Akademikerverlag.

Bae, Suk J.; Kvam, Paul H. (2004): A Nonlinear Random-Coefficients Model for Degradation Testing. In *Technometrics* 46 (4), pp. 460–469.

Bagheri, Behrad; Siegel, David; Zhao, Wenyu; Lee, Jay (2015): A Stochastic Asset Life Prediction Method for Large Fleet Datasets in Big Data Environment. In : Proceedings of ASME International Mechanical Engineering Congress and Exposition. Houston, United States.

Becker, Jörg; Knackstedt, Ralf (2003): Konstruktion und Anwendung fachkonzeptioneller Referenzmodelle im Data Warehousing. In : Wirtschaftsinformatik 2003, vol. 31. 2nd ed. Heidelberg: Physica-Verlag Heidelberg, pp. 415–434.

Becker, Jörg; Knackstedt, Ralf; Pfeiffer, Daniel; Janiesch, Christian (2007): Configurative Method Engineering. On the Applicability of Reference Modeling Mechanisms in Method Engineering. In : Proceedings of Americas Conference on Information Systems. Keystone, United States, pp. 1–12.

Becker, Jörg; Kugeler, Martin; Rosemann, Michael (Eds.) (2011): Process Management. A Guide for the Design of Business Processes. 2nd ed. Berlin: Springer.

Bektas, Oguz; Jones, Jeffrey A.; Sankararaman, Shankar; Roychoudhury, Indranil; Goebel, Kai (2019): A Neural Network Filtering Approach for Similarity-Based Remaining Useful Life Estimation. In *The International Journal of Advanced Manufacturing Technology* 101 (1-4), pp. 87–103.

Bengtsson, Marcus (2004): Condition Based Maintenance Systems. An Investigation of Technical Constituents and Organizational Aspects. Mälardalen University Eskilstuna, Västerås, Sweden.

Bengtsson, Marcus; Olsson, Erik; Funk, Peter; Jackson, Mats (2004): Technical Design of Condition Based Maintenance Systems. A Case Study Using Sound Analysis and Case-Based Reasoning. In : Proceedings of International Conference of Maintenance and Realiability. Knowville, United States, p. 57.

Benkedjouh, Tarak; Medjaher, Kamal; Zerhouni, Noureddine; Rechak, Said (2013): Remaining Useful Life Estimation Based on Nonlinear Feature Reduction and Support Vector Regression. In *Engineering Applications of Artificial Intelligence* 26 (7), pp. 1751–1760.

Bey-Temsamani, Abdellatif; Engels, Marc; Motten, Andy; Vandenplas, Steve; Ompusunggu, Agusmian P. (2009): A Practical Approach to Combine Data Mining and Prognostics for Improved Predictive Maintenance. In : Proceedings of International Workshop on Data Mining Case Studies, vol. 3. Paris, France, pp. 36–43.

Biedermann, Hubert (2008): Ersatzteilmanagement. Effiziente Ersatzteillogistik für Industrieunternehmen. 2., erw. und aktualisierte Auflage. Berlin, Germany: Springer.

Bishop, Christopher M. (2009): Pattern Recognition and Machine Learning. 8th ed. New York, United States: Springer.

Blanca, María J.; Alarcón, Rafael; Arnau, Jaume; Bono, Roser; Bendayan, Rebecca (2017): Non-Normal Data: Is ANOVA Still a Valid Option? In *Psicothema* 29 (4), pp. 552–557.

Bleakie, Alexander; Djurdjanovic, Dragan (2013): Analytical Approach to Similarity-Based Prediction of Manufacturing System Performance. In *Computers in Industry* 64 (6), pp. 625–633.

Bloch, Heinz P.; Geitner, Fred K. (2012): Machinery Failure Analysis and Troubleshooting. Practical Machinery Management for Process Plants. 4th ed. Oxford, United Kingdom: Butterworth-Heinemann.

Blume, Marcel; Allgeyer, Thomas, Giridhar, Sriram; Spielbauer, Stefan (2017): Customers' Voice: Predictive Maintenance in Manufacturing, Western Europe. Status Quo, Approach, Customer Needs, Decision Paths. Frenus GmbH. Stuttgart, 2017.

Bonissone, Piero P. (2006): Knowledge and Time: A Framework for Soft Computing Applications in Prognostics and Health Management (PHM). In : Proceedings of IPMU.

Bonissone, Piero P.; Goebel, Kai (2002): When will it break? A Hybrid Soft Computing Model to Predict Time-to-break Margins in Paper Machines. In : Proceedings of International Symposium on Optical Science and Technology. Seattle, United States, pp. 53–64.

Bonissone, Piero P.; Iyer, Naresh (2007): Soft Computing Applications to Prognostics and Health Management (PHM): Leveraging Field Data and Domain Knowledge. In : Proceedings of International Work-Conference on Artificial Neural Networks, vol. 4507. San Sebastián, Spain. Berlin, Heidelberg, Germany: Springer (4507), pp. 928–939.

Bonissone, Piero P.; Varma, Anil (2005): Predicting the Best Units Within a Fleet: Prognostic Capabilities Enabled by Peer Learning, Fuzzy Similarity, and Evolutionary Design Process. In : Proceedings of International Conference on Fuzzy Systems. Reno, United States, pp. 312–318.

Bonissone, Piero P.; Varma, Anil; Aggour, Kareem S. (2005): A Fuzzy Instance-Based Model for Predicting Expected Life: A Locomotive Application. In : Proceedings of International Conference on Computational Intelligence for Measurement Systems and Applications. Messian, Italy, pp. 20–25.

Bougacha, Omar; Varnier, Christophe; Zerhouni, Noureddine (2020): Enhancing Decisions in Prognostics and Health Management Framework. In *International Journal of Prognostics and Health Management*.

Bousdekis, Alexandros; Magoutas, Babis; Apostolou, Dimitris; Mentzas, Gregoris (2015a): A Proactive Decision-Making Framework for Condition-Based Maintenance. In *Industrial Management & Data Systems* 115 (7), pp. 1225–1250.

Bousdekis, Alexandros; Magoutas, Babis; Apostolou, Dimitris; Mentzas, Gregoris (2015b): Supporting the Selection of Prognostic-Based Decision Support Methods in Manufacturing. In :

Proceedings of International Conference on Enterprise Information Systems. Barcelona, Spain, pp. 487–494.

Bousdekis, Alexandros; Magoutas, Babis; Apostolou, Dimitris; Mentzas, Gregoris (2018): Review, Analysis and Synthesis of Prognostic-Based Decision Support Methods for Condition Based Maintenance. In *Journal of Intelligent Manufacturing* 29 (6), pp. 1303–1316.

Box, George E. P.; Jenkins, Gwilym M.; Reinsel, Gregory C.; Ljung, Greta M. (2016): Time Series Analysis. Forecasting and Control. 5th ed. Hoboken, United States: John Wiley & Sons Inc.

Braun, Virginia; Clarke, Victoria; Hayfield, Nikki; Terry, Gareth (2019): Thematic Analysis. In Pranee Liamputtong (Ed.): Handbook of Research Methods in Health Social Sciences, vol. 3. Singapore: Springer Singapore, pp. 843–860.

Breiman, Leo (2001): Statistical Modeling: The Two Cultures. In *Statistical Science* 16 (3), pp. 199–231.

Breuel, Thomas M. (2015): The Effects of Hyperparameters on SGD Training of Neural Networks. Available online at http://arxiv.org/pdf/1508.02788v1.

Byttner, S.; Rögnvaldsson, T.; Svensson, M. (2011): Consensus Self-Organized Models for Fault Detection. In *Engineering Applications of Artificial Intelligence* 24 (5), pp. 833–839.

Bzdok, Danilo; Altman, Naomi; Krzywinski, Martin (2018): Statistics Versus Machine Learning. In *Nature methods* 15 (4), pp. 233–234.

Cambridge Dictionary: Definition of Approach. Available online at https://dictionary.cambridge.org/de/worterbuch/englisch/approach, checked on 4/8/2021.

Cambridge Dictionary: Definition of Method. Available online at https://dictionary.cambridge.org/de/worterbuch/englisch/method, checked on 4/8/2021.

Carnero Moya, M. C. (2004): The Control of the Setting up of a Predictive Maintenance Programme Using a System of Indicators. In *Omega* 32 (1), pp. 57–75.

Chang, Moon-Hwan; Kang, Myeongsu; Pecht, Michael (2017): Prognostics-Based LED Qualification Using Similarity-Based Statistical Measure with RVM Regression Model. In *IEEE Transactions on Industrial Electronics* 64 (7), pp. 5667–5677.

Chapman, Pete; Clinton, Julian; Kerber, Randy; Khabaza, Thomas; Reinartz, Thomas; Shearer, Colin; Wirth, Rudiger (2000): CRISP-DM 1.0. Step-by-Step Data Mining Guide.

Chebel-Morello, Brigitte; Varnier, Christophe; Nicod, Jean-Marc (2017): From Prognostics and Health Systems Management to Predictive Maintenance 2. Knowledge, Traceability and Decision. Hoboken, United States, London, United Kingdom: John Wiley & Sons Inc; ISTE Ltd (7).

Chen, Lei; Özsu, M. T.; Oria, Vincent (2005): Robust and Fast Similarity Search for Moving Object Trajectories. In : Proceedings of International Conference on Management of Data. Baltimore, Maryland, p. 491.

Cherdantseva, Yulia; Hilton, Jeremy; Rana, Omer; Ivins, Wendy (2016): A Multifaceted Evaluation of the Reference Model of Information Assurance & Security. In *Computers & Security* 63, pp. 45–66.

Coble, Jamie B. (2010): Merging Data Sources to Predict Remaining Useful Life - An Automated Method to Identify Prognostic Parameters. University of Tennessee, Knoxville, United States.

Coble, Jamie B.; Hines Wesley, J. (2008): Prognostic Algorithm Categorization with PHM Challenge Application. In : Proceedings of International Conference on Prognostics and Health Management. Denver, United States, pp. 1–11.

Coble, Jamie B.; Hines Wesley, J. (2011): Applying the General Path Model to Estimation of Remaining Useful Life. In *International Journal of Prognostics and Health Management* 2 (1), pp. 71–82.

Cocheteux, Pierre; Voisin, Alexandre; Levrat, Eric; Iung, Benoît (2009): Prognostic Design: Requirements and Tools. In : Proceedings of International Conference on The Modern Information Technology. Bergame, Italy.

ISO 13374, 2003: Condition Monitoring and Diagnostics of Machines - Data Processing, Communication and Presentation, Part 1: General Guidelines.

ISO 17359, 2018: Condition Monitoring and Diagnostics of Machines - General Guidelines.

ISO 13381, 2015: Condition Monitoring and Diagnostics of Machines - Prognostics, Part 1: General Guidelines.

ISO 13379, 2012: Condition Monitoring and Diagnostics of Machines -Data Interpretation and Diagnostics Techniques, Part 1: General Guidelines.

Cooper, Harris M. (1988): Organizing Knowledge Syntheses: A Taxonomy of Literature Reviews. In *Knowledge in Society* 1 (1), pp. 104–126.

Cristaldi, Loredana; Leone, Giacomo; Ottoboni, Roberto; Subbiah, Subanatarajan; Turrin, Simone (2016): A Comparative Study on Data-Driven Prognostic Approaches Using Fleet Knowledge. In : Proceedings of International Instrumentation and Measurement Technology Conference, pp. 1–6.

Das, Sreerupa (2015): An Efficient Way to Enable Prognostics in an Onboard System. In : Proceedings of IEEE Aerospace Conference. Big Sky, United States, pp. 1–7.

Davenport, Thomas H. (1993): Process Innovation. Reengineering Work Through Information Technology. Boston, United States: Harvard Business School Press.

Davies, Chris; Greenough, R. M. (2000): The Use of Information Systems in Fault Diagnosis. In : Proceedings of Advanced in Manufacturing Technology Conference, vol. 14, pp. 383–388.

Di Maio, Francesco; Zio, Enrico (2013): Failure Prognostics by a Data-Driven Similarity-Based Approach. In *International Journal of Reliability, Quality and Safety Engineering* 20, p. 1350001.

Dragomir, Otilia Elena; Gourtveau, Rafael; Zerhouni, Noureddine; Dragomir, Florin (2007): Framework for a Distributed and Hybrid Prognostic. In : Proceedings of International Federation of Automatic Control Conference on Management and Control of Production and Logistics, vol. 4. Sibiu, Romania, pp. 431–436.

Dul, Jan; Hak, Tony (2008): Case Study Methodology in Business Research. Oxford: Butterworth-Heinemann.

Duong, Pham Luu Trung; Park, Hyunseok; Raghavan, Nagarajan (2018): Application of Multi-Output Gaussian Process Regression for Remaining Useful Life Prediction of Light Emitting Diodes. In *Microelectronics Reliability* 88-90, pp. 80–84.

Duong, Pham Luu Trung; Raghavan, Nagarajan (2018): Prognostic Health Management for LED with Missing Data: Multi-task Gaussian Process Regression Approach. In : Proceedings of Prognostics and System Health Management Conference. Chongqing, China, pp. 1182–1187.

Eker, O. F.; Camci, Fatih; Jennions, I. K. (2014): A Similarity-Based Prognostics Approach for Remaining Useful Life Prediction. In : Proceedings of European Conference of the Prognostics and Health Management Society. Nantes, France, pp. 1–5.

Elattar, Hatem M.; Elminir, Hamdy K.; Riad, A. M. (2016): Prognostics: A Literature Review. In *Complex and Intelligent Systems* (2), pp. 125–154.

Fan, Yuantao; Nowaczyk, Sławomir; Rögnvaldsson, Thorsteinn (2015): Evaluation of Self-Organized Approach for Predicting Compressor Faults in a City Bus Fleet. In *Procedia Computer Science* 53, pp. 447–456.

Fang, Bai; Hongfu, Zuo; Shuhong, Ren (2010): Average Life Prediction for Aero-Engine Fleet Based on Performance Degradation Data. In : Proceedings of Prognostics and System Health Management Conference. Macao, China, pp. 1–6.

Fayyad, Usama; Piatetsky-Shapiro, Gregory; Smyth, Padhraic (1996): Knowledge Discovery and Data Mining. Towards a Unifying Framework. In : Proceedings of International Conference on Knowledge Discovery and Data Mining, vol. 2. Portland, United States, pp. 82–88.

Feldmann, Sebastian; Herweg, Oliver; Rauen, Hartmut; Synek, Peter-Michael (2017): Predictive Maintenance. Service der Zukunft - Und wo er wirklich steht. Edited by Roland Berge GmbH.

Fettke, Peter; Loos, Peter (2003): Multiperspective Evaluation of Reference Models - Towards a Framework. In : Proceedings of Conceptual Modeling for Novel Application Domains. Berlin, Heidelberg, Germany, pp. 80–91.

Fettke, Peter; Loos, Peter; Zwicker, Jörg (2006): Business Process Reference Models: Survey and Classification. In Christoph Bussler, Armin Haller, et al. (Eds.): Business Process Management Workshops, vol. 3812. Berlin, Germany: Springer (3812), pp. 469–483.

Fleischmann, Albert; Oppl, Stefan; Schmidt, Werner; Stary, Christian (2018): Ganzheitliche Digitalisierung von Prozessen. Wiesbaden: Springer Fachmedien Wiesbaden.

Frank, Ulrich (2007): Evaluation of Reference Models. In Peter Fettke, Peter Loos (Eds.): Reference Modeling for Business Systems Analysis: Idea Group Publishing.

Frederick, Dean K.; DeCastro, Jonathan A.; Litt, Jonathan S. (2007): User's Guide for the Commercial Modular Aero-Propulsion System Simulation (C-MAPSS).

Frisk, Erik; Krysander, Mattias (2015): Treatment of Accumulative Variables in Data-Driven Prognostics of Lead-Acid Batteries. In *IFAC-PapersOnLine* 48 (21), pp. 105–112.

Funk, Peter; Jackson, Mats (2005): Experience Based Diagnostics and Condition Based Maintenance Within Production Systems. In : Proceedings of International Congress and Exhibition on Condition Monitoring and Diagnostic Engineering Management, vol. 18. Cranfield, United Kingdom, p. 7.

Gadatsch, Andreas (2017): Grundkurs Geschäftsprozess-Management. Analyse, Modellierung, Optimierung und Controlling von Prozessen. 8., vollständig überarbeitete Auflage. Wiesbaden: Springer Vieweg.

Galliers, Robert D.: Choosing Appropriate Information Systems Research Approaches: A Revised Taxonomy. In : Information Systems Research: Contemporary Approaches and Emergent Traditions. Amsterdam, Netherlands, pp. 327–345.

Gebraeel, Nagi Z. (2006): Sensory-Updated Residual Life Distributions for Components with Exponential Degradation Patterns. In *IEEE Transactions on Automation Science and Engineering* 3 (4), pp. 382–393.

Gebraeel, Nagi Z. (2010): Prognostics-Based Identification of the Top-k Units in a Fleet. In *IEEE Transactions on Automation Science and Engineering* 7 (1), pp. 37–48.

Gebraeel, Nagi Z.; Lawley, M.; Liu, R.; Parmeshwaran, V. (2004): Residual Life Predictions from Vibration-Based Degradation Signals. A Neural Network Approach. In *IEEE Transactions on Industrial Electronics* 51 (3), pp. 694–700.

Gebraeel, Nagi Z.; Lawley, Mark A. (2008): A Neural Network Degradation Model for Computing and Updating Residual Life Distributions. In *IEEE Transactions on Automation Science and Engineering* 5 (1), pp. 154–163.

Gebraeel, Nagi Z.; Lawley, Mark A.; Li, Rong; Ryan, Jennifer K. (2005): Residual-Life Distributions from Component Degradation Signals: A Bayesian Approach. In *IIE Transactions* 37 (6), pp. 543–557.

Ghahramani, Zoubin (2001): An Introduction to Hidden Markov Models and Bayesian Networks. In *International Journal of Pattern Recognition and Artificial Intelligence* 15 (1), pp. 9–42.

Ghodrati, Behzad (2005): Reliability and Operating Environment Based Spare Parts Planning. Luleå University of Technology, Luleå, Sweden.

Glass, Gene V.; Peckham, Percy D.; Sanders, James R. (1972): Consequences of Failure to Meet Assumptions Underlying the Fixed Effects Analyses of Variance and Covariance. In *Review of Educational Research* 42 (3), pp. 237–288.

Goebel, Kai; Daigle, Matthew; Saxena, Abhinav; Sankararaman, Shankar; Roychoudhury, Indranil; Celaya, José R. (2017): Prognostics. The Science of Prediction: CreateSpace Independent Publishing Platform.

Goebel, Kai; Saha, Bhaskar; Saxena, Abhinav (2008): A Comparison of Three Data-Driven Techniques for Prognostics. In : Proceedings of Society for Machinery Failure Prevention Technology, vol. 62. Virginia, United States, pp. 119–131.

Goodfellow, Ian; Bengio, Yoshua; Courville, Aaron (2016): Deep Learning: MIT Press.

Goodman, Douglas; Hofmeister, James P.; Szidarovszky, Ferenc (2019): Prognostics and Health Management. A Practical Approach to Improving System Reliability Using Condition-Based Data: Wiley.

Gouriveau, Rafael; Medjaher, Kamal; Zerhouni, Noureddine (2016): From Prognostics and Health Systems Management to Predictive Maintenance 1. Monitoring and Prognostics. London, United Kingdom, Hoboken, United States: ISTE; Wiley.

Grant, Maria J.; Booth, Andrew (2009): A Typology of Reviews: An Analysis of 14 Review Types and Associated Methodologies. In *Health information and libraries journal* 26 (2), pp. 91–108.

Greff, Klaus; Srivastava, Rupesh Kumar; Koutník, Jan; Steunebrink, Bas R.; Schmidhuber, Jürgen (2017): LSTM: A Search Space Odyssey. In *IEEE transactions on neural networks and learning systems* 28 (10), pp. 2222–2232.

DIN 31051, 2012: Grundlagen der Instandhaltung.

Guepie, Blaise K.; Lecoeuche, Stephane (2015): Similarity-Based Residual Useful Life Prediction for Partially Unknown Cycle Varying Degradation. In : Proceedings of International Conference on Prognostics and Health Management. Austin, United States, pp. 1–7.

ISO 21500:2012: Guidance on Project Management.

Guillén, Antonio J.; Crespo, A.; Macchi, M.; Gómez, J. (2016a): On the Role of Prognostics and Health Management in Advanced Maintenance Systems. In *Production Planning & Control* 27 (12), pp. 991–1004.

Guillén, Antonio J.; González-Prida, Vicente; Gómez, Juan Fco; Crespo, Adolfo (2016b): Standards as Reference to Build a PHM-Based Solution. In : Proceedings of World Congress on Engineering Asset Management, vol. 10. Tampere, Finland, pp. 207–214.

Gupta, J.; Trinquier, Christian; Lorton, Ariane; Feuillard, Vincent (2012): Characterization of Prognosis Methods: An Industrial Approach. In : Proceedings of European Conference of the Prognostics and Health Management Society. Dresden, Germany, pp. 1–9.

Haarman, Mark; Klerk, Pieter de; Decaigny, Peter; Mulders, Michel; Vassiliadis, Costas; Sijsema, Hedwich; Gallo, Ivan (2018): Predictive Maintenance 4.0. Beyond the Hype: PdM 4.0 Delivers Results. Edited by PricewaterhouseCoopers B.V.

Haddad, Gilbert; Sandborn, Peter A.; Pecht, Michael G. (2012): An Options Approach for Decision Support of Systems with Prognostic Capabilities. In *IEEE Transactions on Reliability* 61 (4), pp. 872–883.

Hall, David L.; Llinas, James (1997): An Introduction to Multisensor Data Fusion. In *Proceedings of the IEEE* 85 (1), pp. 6–23.

Hallerbach, Alena; Bauer, Thomas; Reichert, Manfred (2008): Managing Process Variants in the Process Lifecycle. In : Proceedings of International Conference on Enterprise Information Systems, vol. 10. Barcelona, Spain.

Han, Jiawei; Kamber, Micheline; Pei, Jian (2012): Data Mining. Concepts and Techniques. 3rd ed. Waltham, United States: Morgan Kaufmann.

Hars, Alexander (1994): Referenzdatenmodelle. Grundlagen effizienter Datenmodellierung. 1st ed. Wiesbaden: Gabler Verlag.

Hart, Douglas (2017): Implementing a "Best Practices" Predictive Maintenance Program. Avoiding the 10 Most Common Pitfalls. Edited by Emerson Reliability Consulting.

Heimes, Felix O. (2008): Recurrent Neural Networks for Remaining Useful Life Estimation. In : Proceedings of International Conference on Prognostics and Health Management. Denver, United States, pp. 1–6.

Hendrickx, Kilian; Meert, Wannes; Cornelis, Bram; Janssens, Karl; Gryllias, Konstantinos; Davis, Jesse (2019): A Fleet-Wide Approach for Condition Monitoring of Similar Machines Using Time-Series Clustering. In Alfonso F. D. Rincon (Ed.): Advances in Condition Monitoring of Machinery in Non-Stationary Operations, vol. 15. Cham, Switzerland: Springer, pp. 101–110.

Heng, Aiwina; Zhang, Sheng; Tan, Andy C.C.; Mathew, Joseph (2009): Rotating Machinery Prognostics. State of the Art, Challenges and Opportunities. In *Mechanical Systems and Signal Processing* 23 (3), pp. 724–739.

Hevner, Alan R. (2007): A Three Cycle View of Design Science Research. In *Scandinavian journal of information systems* 19 (2), p. 4.

Hevner, Alan R.; March, Salvatore T.; Park, Jinsoo; Ram, Sudha (2004): Design Science in Information Systems Research. In *MIS quarterly* 28 (1), pp. 75–105.

Hickey, Ann M.; Davis, Alan M. (2003): Requirements Elicitation and Elicitation Technique Selection. Model for Two Knowledge-Intensive Software Development Processes. In : Proceedings of Hawaii International Conference on System Sciences, vol. 36. Big Island, USA, 1-10.

Hu, Chao; Youn, Byeng D.; Wang, Pingfeng; Taek Yoon, Joung (2012): Ensemble of Data-Driven Prognostic Algorithms for Robust Prediction of Remaining Useful Life. In *Reliability Engineering and System Safety* 103, pp. 120–135.

Huang, Bin; Di, Yuan; Jin, Chao; Lee, Jay (2017): Review of Data-Driven Prognostics and Health Management Techniques. Lessons Learned from PHM Data Challenge Competitions. In : Proceeding of Machine Failure Prevention Technology. Virginia Beach, United States, pp. 1–17.

Huang, Hong-Zhong; Wang, Hai-Kun; Li, Yan-Feng; Zhang, Longlong; Liu, Zhiliang (2015): Support Vector Machine Based Estimation of Remaining Useful Life. Current Research Status and Future Trends. In *Journal of Mechanical Science and Technology* 29 (1), pp. 151–163.

Huang, Runqing; Xi, Lifeng; Li, Xinglin; Liu, C. R.; Qiu, Hai; Lee, Jay (2007): Residual Life Predictions for Ball Bearings Based on Self-Organizing Map and Back Propagation Neural Network Methods. In *Mechanical Systems and Signal Processing* 21 (1), pp. 193–207.

Huiskonen, Janne (2001): Maintenance Spare Parts Logistics. Special Characteristics and Strategic Choices. In *International Journal of Production Economics* 71 (1), pp. 125–133.

IEEE Std 1856-2017: IEEE Standard Framework for Prognostics and Health Management of Electronic Systems.

ISO 5807, 1985: Information Processing — Documentation Symbols and Conventions for Data, Program and System Flowcharts, Program Network Charts and System Resources Charts.

DIN EN 13306, 2010-12: Instandhaltung.

Jahani, Alireza; Akhavan, Peyman; Jafari, Mostafa; Fathian, Mohammad (2016): Conceptual Model for Knowledge Discovery Process in Databases Based on Multi-Agent System. In *VINE Journal of Information and Knowledge Management Systems* 46 (2), pp. 207–231.

Jardine, Andrew K.S.; Lin, Daming; Banjevic, Dragan (2006): A Review on Machinery Diagnostics and Prognostics Implementing Condition-Based Maintenance. In *Mechanical Systems and Signal Processing* 20 (7), pp. 1483–1510.

Jia, Xiaodong; Huang, Bin; Feng, Jianshe; Cai, Haoshu; Lee, Jay (2018): A Review of PHM Data Competitions from 2008 to 2017. In *PHM Society Conference* 10 (1).

Jin, Chao; Djurdjanovic, Dragan; Ardakani, Hossein D.; Wang, Keren; Buzza, Matthew; Begheri, Behrad et al. (2015): A Comprehensive Framework of Factory-to-Factory Dynamic Fleet-Level Prognostics and Operation Management for Geographically Distributed Assets. In : Proceedings of IEEE International Conference on Automation Science and Engineering. Gothenburg, Sweden, pp. 225–230.

Johnson, Preston (2012): Fleet Wide Asset Monitoring. Sensory Data to Signal Processing to Prognostics. In : Proceedings of Annual Conference of the Prognostics and Health Management Society. Minneapolis, United States, pp. 23–27.

Kammoun, Mohamed A.; Rezg, Nidhal (2018): Toward the Optimal Selective Maintenance for Multi-Component Systems Using Observed Failure: Applied to the FMS Study Case. In *International Journal of Advanced Manufacturing Technology* 96 (1-4), pp. 1093–1107.

Kan, Man S.; Tan, Andy C.C.; Mathew, Joseph (2015): A Review on Prognostic Techniques for Non-Stationary and Non-Linear Rotating Systems. In *Mechanical Systems and Signal Processing* 62-63, pp. 1–20.

Katipamula, Srinivas; Brambley, Michael (2005): Review Article: Methods for Fault Detection, Diagnostics, and Prognostics for Building Systems - A Review, Part I. In *HVAC&R Research* 11 (1), pp. 3–25.

Kaufman, Leonard; Rousseeuw, Peter J. (2005): Finding Groups in Data. An Introduction to Cluster Analysis. Hoboken, United States: Wiley-Interscience.

KDnuggets (2014): What Main Methodology Are You Using for Your Analytics, Data Mining, or Data Science Projects? Poll. Available online at https://www.kdnuggets.com/polls/2014/analytics-data-mining-data-science-methodology.html, checked on 9/18/2019.

Khazraei, Khashayar; Deuse, Jochen (2011): A Strategic Standpoint on Maintenance Taxonomy. In *Journal of Facilities Management* 9 (2), pp. 96–113.

Khelif, Racha; Chebel-Morello, Brigitte; Malinowski, Simon; Laajili, Emna; Fnaiech, Farhat; Zerhouni, Noureddine (2017): Direct Remaining Useful Life Estimation Based on Support Vector Regression. In *IEEE Transactions on Industrial Electronics* 64 (3), pp. 2276–2285.

Kim, Nam-Ho; An, Dawn; Choi, Joo-Ho (2017): Prognostics and Health Management of Engineering Systems. An Introduction. Cham, Switzerland: Springer International Publishing.

Kodinariya, Trupti; Makwana, P. R. (2013): Review on Determining of Cluster in K-Means Clustering. In *International Journal of Advance Research in Computer Science and Management Studies* 1, pp. 90–95.

Kothamasu, Ranganath; Huang, Samuel H.; VerDuin, William H. (2006): System Health Monitoring and Prognostics - A Review of Current Paradigms and Practices. In *International Journal of Advanced Manufacturing Technology* 28, pp. 1012–1024.

Kotsiantis, Sotiris B.; Zaharakis, I.; Pintelas, P. (2007): Supervised Machine Learning: A Review of Classification Techniques. In *Emerging artificial intelligence applications in computer engineering* 160, pp. 3–24.

Krause, Jakob; Cech, Sebastian; Rosenthal, Frank; Gossling, Andreas; Groba, Christin; Vasyutynskyy, Volodymyr (2010): Factory-Wide Predictive Maintenance in Heterogeneous Environments. In : Proceedings of IEEE International Workshop on Factory Communication Systems, vol. 8. Nancy, France, pp. 153–156.

Kumar, Sachin; Torres, Myra; Chan, Y. C.; Pecht, Michael (2008): A Hybrid Prognostics Methodology for Electronic Products. In : Proceedings of IEEE International Joint Conference on Neural Networks. Hong Kong, China, pp. 3479–3485.

Kvale, Steinar; Brinkmann, Svend (2009): Interviews. Learning the Craft of Qualitative Research Interviewing. 2nd ed. Los Angeles, London, New Delhi, Singapore: USAGE.

Lam, Jack; Sankararaman, Shankar; Stewart, Bryan (2014): Enhanced Trajectory Based Similarity Prediction with Uncertainty Quantification. In : Proceedings of Annual Conference of the Prognostics and Health Management Society. Fort Worth, United States.

Lamoureux, Benjamin; Masse, Jean-Remi; Mechbal, Nazih (2015): Towards an Integrated Development of PHM Systems for Aircraft Engines: In-Design Selection and Validation of Health Indicators. In : Proceedings of IEEE Conference on Prognostics and Health Management. Austin, United States, pp. 1–8.

Lapira, Edzel (2012): Fault Detection in a Network of Similar Machines Using Clustering Approach. University of Cincinnati, Cincinnati, United States.

Le, Tung; Geramifard, Omid (2014): Fleet-Based Approach for Tool Wear Estimation Using Sequential Importance Sampling with Resampling. In : Proceedings of International Conference on Control, Automation, Robotics & Vision, vol. 13. Singapore, pp. 1467–1472.

Lebold, Mitchell; Reichard, Karl; Boylan, David (2003): Utilizing DCOM in an Open System Architecture Framework for Machinery Monitoring and Diagnostics. In : Proceedings of IEEE Aerospace Conference. Big Sky, United States, 3_1227-3_1236.

LeCun, Yann; Bengio, Yoshua; Hinton, Geoffrey (2015): Deep Learning. In *Nature* 521 (7553), pp. 436–444.

Lee, Jay; Bagheri, Behrad (2015): Cyber-Physical Systems in Future Maintenance. In Joe Amadi-Echendu, Changela Hoohlo, Joe Mathew (Eds.): 9th WCEAM Research Papers: Springer International Publishing, pp. 299–305.

Lee, Jay; Bagheri, Behrad; Kao, Hung-An (2015): A Cyber-Physical Systems Architecture for Industry 4.0-Based Manufacturing Systems. In *Manufacturing Letters* 3, pp. 18–23.

Lee, Jay; Chen, Yan; Al-Atat, Hassan; AbuAli, Mohamed; Lapira, Edzel (2009a): A Systematic Approach for Predictive Maintenance Service Design. Methodology and Applications. In *International Journal of Internet Manufacturing and Services* 2 (1), p. 76.

Lee, Jay; Jin, Chao; Liu, Zongchang; Davari Ardakani, Hossein (2017): Introduction to Data-Driven Methodologies for Prognostics and Health Management. In Stephen Ekwaro-Osire, Aparecido C. Gonçalves, Fisseha M. Alemayehu (Eds.): Probabilistic Prognostics and Health Management of Energy Systems. Cham, Switzerland: Springer International Publishing, pp. 9–32.

Lee, Jay; Kao, Hung-An; Yang, Shanhu (2014a): Service Innovation and Smart Analytics for Industry 4.0 and Big Data Environment. In *Procedia CIRP* 16, pp. 3–8.

Lee, Jay; Liao, Linxia; Lapira, Edzel; Ni, Jun; Li, Lin (2009b): Informatics Platform for Designing and Deploying e-Manufacturing Systems. In Lihui Wang, Andrew Y.C Nee (Eds.): Collaborative Design and Planning for Digital Manufacturing. Online-Ausgabe. London, United Kingdom: Springer London, pp. 1–35.

Lee, Jay; Wang, Ben (1999): Computer-Aided Maintenance. Methodologies and Practices. Boston, United States: Springer.

Lee, Jay; Wu, Fangji; Zhao, Wenyu; Ghaffari, Masoud; Liao, Linxia; Siegel, David (2014b): Prognostics and Health Management Design for Rotary Machinery Systems - Reviews, Methodology and Applications. In *Mechanical Systems and Signal Processing* 42 (1-2), pp. 314–334.

Léger, Jean-Baptiste; Iung, Benoît (2012): Ships Fleet-Wide Management and Naval Mission Prognostics: Lessons Learned and New Issues. In : Proceedings of IEEE Conference on Prognostics and Health Management. Denver, United States.

Léger, Jean-Baptiste; Morel, Gérard (2001): Integration of Maintenance in the Enterprise: Towards an Enterprise Modelling-Based Framework Compliant with Proactive Maintenance Strategy. In *Production Planning & Control* 12 (2), pp. 176–187.

Lei, Yaguo; Guo, Liang; Li, Naipeng; Yan, Tao (2018): Machinery Health Prognostics: A Systematic Review from Data Acquisition to RUL Prediction. In *Mechanical Systems and Signal Processing* 104, pp. 799–834.

Leone, Giacomo; Cristaldi, Loredana; Turrin, Simone (2017): A Data-Driven Prognostic Approach Based on Statistical Similarity: An Application to Industrial Circuit Breakers. In *Measurement*, pp. 163–170.

Levrat, Eric; Iung, Benoît; Crespo Márquez, Adolfo (2008): E-Maintenance: Review and Conceptual Framework. In *Production Planning & Control* 19 (4), pp. 408–429.

Li, Lisha; Jamieson, Kevin; DeSalvo, Giulia; Rostamizadeh, Afshin; Talwalkar, Ameet (2018a): Hyperband: A Novel Bandit-Based Approach to Hyperparameter Optimization. In *Journal of Machine Learning Research* 18, pp. 1–52.

Li, Rui; Verhagen, Wim J.C.; Curran, Richard (2018b): A Comparative Study of Data-Driven Prognostic Approaches: Stochastic and Statistical Models. In : Proceedings of IEEE International Conference on Prognostics and Health Management. Seattle, United States, pp. 1–8.

Li, Rui; Verhagen, Wim J.C.; Curran, Richard (2018c): A Functional Architecture of Prognostics and Health Management Using a Systems Engineering Approach. In : Proceedings of the European Conference of the Prognostics and Health Management Society, vol. 4. Utrecht, Netherlands, pp. 1–10.

Li, Xiaobin; Qian, Jiansheng; Wang, Gai-ge (2015): Fault Prognostic Based on Hybrid Method of State Judgment and Regression. In *Advances in Mechanical Engineering* 5 (7), p. 149562.

Liao, Linxia; Lee, Jay (2009): A Novel Method for Machine Performance Degradation Assessment Based on Fixed Cycle Features Test. In *Journal of Sound and Vibration* 326 (3-5), pp. 894–908.

Liao, T. Warren (2005): Clustering of Time Series Data - A Survey. In *Pattern Recognition* 38 (11), pp. 1857–1874.

Lim, Pin; Goh, Chi K.; Tan, Kay C.; Dutta, Partha (2014): Estimation of Remaining Useful Life Based on Switching Kalman Filter Neural Network Ensemble. In : Proceedings of Annual Conference of the Prognostics and Health Management Society. Fort Worth, United States.

Ling, You; Mahadevan, Sankaran (2012): Integration of Structural Health Monitoring and Fatigue Damage Prognosis. In *Mechanical Systems and Signal Processing* 28, pp. 89–104.

Lipp, Ulrich; Will, Hermann (2004): Das große Workshop-Buch. Konzeption, Inszenierung und Moderation von Klausuren, Besprechungen und Seminaren. 7., aktualisierte und neu ausgestattete Auflage. Weinheim: Beltz.

Liu, Fei T.; Ting, Kai M.; Zhou, Zhi-Hua (2008): Isolation Forest. In : Proceedings of IEEE International Conference on Data Mining, vol. 8. Pisa, Italy, pp. 413–422.

Liu, Jianbo; Djurdjanovic, Dragan; Ni, Jun; Casoetto, Nicolas; Lee, Jay (2007): Similarity Based Method for Manufacturing Process Performance Prediction and Diagnosis. In *Computers in Industry* 58 (6), pp. 558–566.

Liu, Jie; Zio, Enrico (2016): A Framework for Asset Prognostics from Fleet Data. In : Proceedings of Prognostics and System Health Management Conference. Chengdu, China, pp. 1–5.

Lix, Lisa M.; Keselman, Joanne C.; Keselman, H. J. (1996): Consequences of Assumption Violations Revisited: A Quantitative Review of Alternatives to the One-Way Analysis of Variance F Test. In *Review of Educational Research* 66 (4), pp. 579–619.

Lobe, Christopher: Bewertung der Eignung von Modellierungssprachen zur ebenenübergreifenden Prozessdarstellung im Managementhandbuch. Otto von Guericke Universität Magdeburg, Magdeburg, Germany.

Lucks, Kai (Ed.) (2017): Praxishandbuch Industrie 4.0. Branchen - Unternehmen - M&A. Stuttgart, Germany: Schäffer-Poeschel Verlag Stuttgart.

March, Salvatore T.; Smith, Gerald F. (1995): Design and Natural Science Research on Information Technology. In *Decision Support Systems* 15 (4), pp. 251–266.

Mariscal, Gonzalo; Marbán, Óscar; Fernández, Covadonga (2010): A Survey of Data Mining and Knowledge Discovery Process Models and Methodologies. In *The Knowledge Engineering Review* 25 (2), pp. 137–166.

Markus, M. L.; Majchrzak, Ann; Gasser, Les (2002): A Design Theory for Systems That Support Emergent Knowledge Processes. In *MIS quarterly* 26 (3), pp. 179–212.

Mathew, Vimala; Toby, Tom; Singh, Vikram; Rao, B. M.; Kumar, M. G. (2017): Prediction of Remaining Useful Lifetime (RUL) of Turbofan Engine Using Machine Learning. In : Proceedings of IEEE International Conference on Circuits and Systems Conference. Thiruvananthapuram, India, pp. 306–311.

Matook, Sabine; Indulska, Marta (2009): Improving the Quality of Process Reference Models: A Quality Function Deployment-Based Approach. In *Decision Support Systems* 47 (1), pp. 60–71.

Medina-Oliva, Gabriela; Voisin, Alexandre; Monnin, Maxime; Léger, Jean-Baptiste (2014): Predictive Diagnosis Based on a Fleet-Wide Ontology Approach. In *Knowledge-Based Systems* 68, pp. 40–57.

Medina-Oliva, Gabriela; Voisin, Alexandre; Monnin, Maxime; Peysson, Flavien; Léger, Jean-Baptiste (2012): Prognostics Assessment Using Fleet-Wide Ontology. In : Proceedings of Annual Conference of the Prognostics and Health Management Society. Minneapolis, United States.

Merriam-Webster Dictionary: Definition of ALGORITHM. Available online at https://www.merriam-webster.com/dictionary/algorithm, checked on 4/8/2021.

Merriam-Webster Dictionary: Definition of FLEET. Available online at https://www.merriam-webster.com/dictionary/fleet, checked on 5/18/2017.

Merriam-Webster Dictionary: Definition of TECHNIQUE. Available online at https://www.merriam-webster.com/dictionary/technique, checked on 4/8/2021.

Michau, Gabriel; Fink, Olga (2019): Unsupervised Fault Detection in Varying Operating Conditions. In : Proceedings of IEEE International Conference on Prognostics and Health Management. San Francisco, United States, pp. 1–10.

Michau, Gabriel; Palmé, Thomas; Fink, Olga (2018): Fleet PHM for Critical Systems: Bi-Level Deep Learning Approach for Fault Detection. In : Proceedings of the European Conference of the Prognostics and Health Management Society, vol. 4. Utrecht, Netherlands.

Millar, Richard C. (2007): A Systems Engineering Approach to PHM for Military Aircraft Propulsion Systems. In : Proceedings of IEEE Aerospace Conference. Big Sky, United States, pp. 1–9.

MIMOSA (1998-2019): Open Standards for Physical Assets. Available online at http://www.mimosa.org/, checked on 10/7/2019.

Mishura, Yuliya; Shevchenko, Georgiy (2017): Theory and Statistical Applications of Stochastic Processes. Hoboken, United States, London: Wiley; ISTE.

Misoch, Sabina (2019): Qualitative Interviews. 2nd ed. Berlin, Germany: De Gruyter Oldenbourg.

Mobley, R. K. (2002): An Introduction to Predictive Maintenance. 2nd ed. Amsterdam, Netherlands: Butterworth-Heinemann.

Monnin, Maxime; Abichou, Bouthaina; Voisin, Alexandre; Mozzati, Christophe (2011a): Fleet Historical Cases for Predictive Maintenance. In : Proceedings of International Conference on Acoustical and Vibratory Methods in Surveillance and Diagnostics. Compiègne, France.

Monnin, Maxime; Voisin, Alexandre; Léger, Jean-Baptiste; Iung, Benoît (2011b): Fleet-Wide Health Management Architecture. In : Proceedings of Annual Conference of the Prognostics and Health Management Society. Montreal, Canada.

Moody, Daniel L.; Shanks, Graeme G. (1994): What Makes a Good Data Model? Evaluating the Quality of Entity Relationship Models. In : Proceedings of International Conference on the Entity-Relationship Approach, vol. 13. Manchester, United Kingdom, pp. 94–111.

Moody, Daniel L.; Shanks, Graeme G. (2003): Improving the Quality of Data Models: Empirical Validation of a Quality Management Framework. In *Information Systems* 28 (6), pp. 619–650.

Mosallam, A.; Medjaher, K.; Zerhouni, N. (2015): Component Based Data-Driven Prognostics for Complex Systems: Methodology and Applications. In : Proceedings of International Conference on Reliability Systems Engineering, vol. 1. Beijing, China, pp. 1–7.

Moubray, John (2001): Reliability-Centered Maintenance: Industrial Press Inc.

Muller, Alexandre; Suhner, Marie-Christine; Iung, Benoît (2008): Formalisation of a New Prognosis Model for Supporting Proactive Maintenance Implementation on Industrial System. In *Reliability Engineering and System Safety* 93 (2), pp. 234–253.

Müller, Andreas C.; Guido, Sarah (2017): Introduction to Machine Learning with Python. A Guide for Data Scientists. 1st ed. Sebastopol, United States: O'Reilly Media.

NASA (2000): Reliability Centered Maintenance Guide for Facilities and Collateral Equipment.

Oxford Dictionaries: Fleet - Definition of Fleet in English. Oxford Dictionaries. Available online at https://en.oxforddictionaries.com/definition/fleet, checked on 5/22/2017.

Palau, Adrià S.; Liang, Zhenglin; Lütgehetmann, Daniel; Parlikad, Ajith K. (2020): Collaborative Prognostics in Social Asset Networks. In Adolfo Crespo Márquez, Marco Macchi, Ajith Parlikad (Eds.): Value Based and Intelligent Asset Management. Mastering the Asset Management Transformation in Industrial Plants and Infrastructures, vol. 37. Cham, Switzerland: Springer, pp. 329–349.

Paré, Guy; Trudel, Marie-Claude; Jaana, Mirou; Kitsiou, Spyros (2015): Synthesizing Information Systems Knowledge: A Typology of Literature Reviews. In *Information & Management* 52 (2), pp. 183–199.

Patrick, Romano; Smith, Matthew J.; Byington, Carl S.; Vachtsevanos, George J.; Tom, Kwok; Ly, Canh (2010): Integrated Software Platform for Fleet Data Analysis, Enhanced Diagnostics, and Safe Transition to Prognostics for Helicopter Component CBM.

Pawellek, G. (2016): Integrierte Instandhaltung und Ersatzteillogistik: Vorgehensweisen, Methoden, Tools: Springer Berlin Heidelberg.

Pecht, Michael (Ed.) (2008): Prognostics and Health Management of Electronics. Hoboken, United States: Wiley.

Pecht, Michael; Jaai, Rubyca (2010): A Prognostics and Health Management Roadmap for Information and Electronics-Rich Systems. In *Microelectronics Reliability* 50 (3), pp. 317–323.

Pecht, Michael; Kumar, Sachin (2008): Data Analysis Approach for System Reliability, Diagnostics and Prognostics. In : Proceedings of Pan Pacific Microelectronics Symposium. Kauai, United States, pp. 22–24.

Peel, Leto (2008): Data Driven Prognostics Using a Kalman Filter Ensemble of Neural Network Models. In : Proceedings of International Conference on Prognostics and Health Management. Denver, United States, pp. 1–6.

Peng, Ying; Dong, Ming; Zuo, Ming J. (2010): Current Status of Machine Prognostics in Condition-Based Maintenance. A Review. In *International Journal of Advanced Manufacturing Technology* 50 (1-4), pp. 297–313.

Peng, Yu; Wang, Hong; Wang, Jianmin; Liu, Datong; Peng, Xiyuan (2012): A Modified Echo State Network Based Remaining Useful Life Estimation Approach. In : Proceedings of IEEE Conference on Prognostics and Health Management. Denver, United States, pp. 1–7.

Project Management Institute (2017): A Guide to the Project Management Body of Knowledge (PMBOK® Guide). 6th ed. Newtown Square, United States: Project Management Institute.

Ramasso, Emmanuel (2014): Investigating Computational Geometry for Failure Prognostics. In *International Journal of Prognostics and Health Management* 5 (1), p. 5.

Ramasso, Emmanuel; Saxena, Abhinav (2014): Performance Benchmarking and Analysis of Prognostic Methods for CMAPSS Datasets.

Rani, Sangeeta; Sikka, Geeta (2012): Recent Techniques of Clustering of Time Series Data: A Survey. In *International Journal of Computer Applications* 52 (15), pp. 1–9.

Rasmussen, Carl E. (2004): Gaussian Processes in Machine Learning: Springer Berlin Heidelberg (3176).

Rasmussen, Carl E.; Williams, Christopher K. I. (2008): Gaussian Processes for Machine Learning. 3rd ed. Cambridge, United States: MIT Press.

Rezvanizaniani, Seyed M.; Liu, Zongchang; Chen, Yan; Lee, Jay (2014): Review and Recent Advances in Battery Health Monitoring and Prognostics Technologies for Electric Vehicle (EV) Safety and Mobility. In *Journal of Power Sources* 256, pp. 110–124.

Riad, A. M.; Elminir, Hamdy K.; Elattar, Hatem M. (2010): Evaluation of Neural Networks in the Subject of Prognostics as Compared to Linear Regression Model. In *International Journal of Engineering & Technology* 10 (6), pp. 52–58.

Rigamonti, Marco; Baraldi, Piero; Zio, Enrico; Roychoudhury, Indranil; Goebel, Kai; Poll, Scott (2015): Echo State Network for the Remaining Useful Life Prediction of a Turbofan Engine. In : Proceedings of Annual Conference of the Prognostics and Health Management Society. Coronado, United States, pp. 255–270.

Rokach, Lior; Maimon, Oded (2005): Clustering Methods. In Oded Maimon, Lior Rokach (Eds.): Data Mining and Knowledge Discovery Handbook. New York, United States: Springer, pp. 321–352.

Rosemann, M.; van der Aalst, W.M.P. (2007): A Configurable Reference Modelling Language. In *Information Systems* 32 (1), pp. 1–23.

Rousseeuw, Peter J. (1987): Silhouettes: A Graphical Aid to the Interpretation and Validation of Cluster Analysis. In *Journal of Computational and Applied Mathematics* 20, pp. 53–65.

Samuel, Arthur L. (1959): Some Studies in Machine Learning Using the Game of Checkers. In *IBM Journal of Research and Development* 3 (3), pp. 210–229.

Santana, Fabiana S.; Costa, Anna H. Reali; Truzzi, Flavio S.; Silva, Felipe L.; Santos, Sheila L.; Francoy, Tiago M.; Saraiva, Antonio M. (2014): A Reference Process for Automating Bee Species Identification Based on Wing Images and Digital Image Processing. In *Ecological Informatics* 24, pp. 248–260.

Santini, Simone; Jain, Ramesh (1999): Similarity Measures. In *IEEE Transactions on Pattern Analysis and Machine Intelligence* 21 (9), pp. 871–883.

Sardá-Espinosa, Alexis (2017): Comparing Time-Series Clustering Algorithms in R Using the dtwclust Package.

Saunders, Sam C. (2007): Reliability, Life Testing and the Prediction of Service Lives. New York, United States: Springer New York.

Saxena, Abhinav; Celaya, José R.; Balaban, Edward; Goebel, Kai; Saha, Bhaskar; Saha, Sankalita; Schwabacher, Mark (2008a): Metrics for Evaluating Performance of Prognostic Techniques. In : Proceedings of International Conference on Prognostics and Health Management. Denver, United States, pp. 1–17.

Saxena, Abhinav; Celaya, José R.; Saha, Bhaskar; Saha, Sankalita; Goebel, Kai (2009): Evaluating Algorithm Performance Metrics Tailored for Prognostics. In : Proceedings of IEEE Aerospace Conference. Big Sky, United States, pp. 1–13.

Saxena, Abhinav; Celaya, José R.; Saha, Bhaskar; Saha, Sankalita; Goebel, Kai (2010a): Metrics for Offline Evaluation of Prognostic Performance. In *International Journal of Prognostics 2010* 1 (1), pp. 4–23.

Saxena, Abhinav; Goebel, Kai (2008): Turbofan Engine Degradation Simulation Data Set. In *NASA Ames Prognostics Data Repository*.

Saxena, Abhinav; Goebel, Kai; Simon, Don; Eklund, Neil (2008b): Damage Propagation Modeling for Aircraft Engine Run-to-Failure Simulation. In : Proceedings of International Conference on Enterprise Information Systems. Barcelona, Spain, pp. 1–9, checked on 4/25/2017.

Saxena, Abhinav; Roychoudhury, Indranil; Celaya, José R.; Saha, Sankalita; Saha, Bhaskar; Goebel, Kai (2010b): Requirements Specifications for Prognostics. An Overview. In *American Institute of Aeronautics and Astronautics*.

Saxena, Abhinav; Sankararaman, Shankar; Goebel, Kai (2014): Performance Evaluation for Fleet-Based and Unit-Based Prognostic Methods.

Saxena, Abhinav; Wu, Biqing; Vachtsevanos, George J. (2005): Integrated Diagnosis and Prognosis Architecture for Fleet Vehicles Using Dynamic Case-Based Reasoning.

Schallmo, Daniel; Williams, Christopher, A.; Boardman, Luke (2017): Digital Transformation of Business Models. Best Practice, Enablers, and Roadmap. In *International Journal of Innovation Management* 21 (08), p. 1740014.

Scheer, August-Wilhelm (1999): „ARIS - House of Business Engineering": Konzept zur Beschreibung und Ausführung von Referenzmodellen. In Jörg Becker, Michael Rosemann, Reinhard Schütte (Eds.): Referenzmodellierung. State-of-the-Art und Entwicklungsperspektiven, vol. 37. Heidelberg, Germany: Physica-Verlag Heidelberg, pp. 2–21.

Schiersmann, Christiane; Thiel, Heinz-Ulrich (2011): Organisationsentwicklung. Prinzipien und Strategien von Veränderungsprozessen. 3., durchgesehene Auflage. Wiesbaden: VS Verlag für Sozialwissenschaften.

Schleichert, Olaf P.; Bringmann, Björn; Zalotskiy, Sergey; Kremer, Hardy; Köpfer, David (2017): Predictive Maintenance. Taking Pro-Active Measures Based on Advanced Data Analytics to Predict and Avoid Machine Failures. Edited by Deloitte Consulting GmbH. Deloitte Analytics Institute.

Schlick, Jochen; Negele, Kathrin (2019): Deutscher Industrie 4.0 Index. Eine Studie der Staufen AG und der Staufen Digital Neonex. Edited by Staufen AG, Digital Neonex GmbH.

Schmider, Emanuel; Ziegler, Matthias; Danay, Erik; Beyer, Luzi; Bühner, Markus (2010): Is It Really Robust? In *Methodology* 6 (4), pp. 147–151.

Schneider, K.; Cassady, C. R. (2004): Fleet Performance Under Selective Maintenance. In : Procedings of Annual Symposium Reliability and Maintainability. Los Angeles, United States, pp. 571–576.

Schütte, Reinhard (1998): Grundsätze ordnungsmäßiger Referenzmodellierung. Wiesbaden, Germany: Gabler Verlag.

Schwegmann, Ansgar (1999): Objektorientierte Referenzmodellierung. Wiesbaden, Germany: Deutscher Universitätsverlag.

Selcuk, Sule (2017): Predictive Maintenance, Its Implementation and Latest Trends. In *Proceedings of the Institution of Mechanical Engineers, Part B: Journal of Engineering Manufacture* 231 (9), pp. 1670–1679.

Shalev-Shwartz, Shai; Ben-David, Shai (2014): Understanding Machine Learning. From Theory to Algorithms. Cambridge, United States: Cambridge University Press.

Shi, Weisong; Dustdar, Schahram (2016): The Promise of Edge Computing. In *Computer* 49 (5), pp. 78–81.

Shi, Zhe; Lee, Jay; Cui, Peng (2016): Prognostics and Health Management Solution Development in LabVIEW: Watchdog agent® Toolkit and Case Study. In : Proceedings of Prognostics and System Health Management Conference. Chengdu, China, pp. 1–6.

Shin, Jong-Ho; Jun, Hong-Bae (2015): On Condition Based Maintenance Policy. In *Journal of Computational Design and Engineering* 2 (2), pp. 119–127.

Shumaker, Brent D.; Ledlow, Jonathan B.; O'Hagan, Ryan D.; McCarter, Dan E.; Hashemian, Hashem M. (2013): Remaining Useful Life Estimation of Electric Cables in Nuclear Power Plants. In *Chemical Engineering Transactions* 33, pp. 877–882.

Si, Xiao-Sheng; Wang, Wenbin; Hu, Chang-Hua; Zhou, Dong-Hua (2011): Remaining Useful Life Estimation - A Review on the Statistical Data Driven Approaches. In *European Journal of Operational Research* 213 (1), pp. 1–14.

Sikorska, Joanna Z.; Hodkiewicz, Melinda; Ma, L. (2011): Prognostic Modelling Options for Remaining Useful Life Estimation by Industry. In *Mechanical Systems and Signal Processing* 25 (5), pp. 1803–1836.

Simon, Herbert A. (1996): The Sciences of the Artificial. 3rd ed. Cambridge, United States: MIT Press.

Singh, Amanpreet; Thakur, Narina; Sharma, Aakanksha (2016): A Review of Supervised Machine Learning Algorithms. In : Proceedings of International Conference on Computing for Sustainable Global Development (INDIACom), vol. 3. New Delhi, India, pp. 1310–1315.

Song, B. L.; Lee, Jay (2013): Framework of Designing an Adaptive and Multi-Regime Prognostics and Health Management for Wind Turbine Reliability and Efficiency Improvement. In *International Journal of Advanced Computer Science and Applications* 4 (2).

Strunz, Matthias (2012): Instandhaltung. Grundlagen - Strategien - Werkstätten. Berlin, Germany: Springer.

Subbiah, Subanatarajan; Turrin, Simone (2015): Extraction and Exploitation of R&M Knowledge from a Fleet Perspective. In : Proceedings of Annual Reliability and Maintainability Symposium (RAMS). Palm Harbor, United States, pp. 1–6.

Sullivan, Greg; Pugh, Ray; Melendez, Aldo P.; Hunt, W. D. (2010): Operations & Maintenance Best Practices. A Guide to Achieving Operational Efficiency. 3rd ed. Pacific Northwest National Laboratory. Richland, United States.

Sun, Bo; Zeng, Shengkui; Kang, Rui; Pecht, Michael (2010): Benefits Analysis of Prognostics in Systems. In : Proceedings of Prognostics and System Health Management Conference. Macao, China, pp. 1–8.

Sun, Bo; Zeng, Shengkui; Kang, Rui; Pecht, Michael G. (2012): Benefits and Challenges of System Prognostics. In *IEEE Transactions on Reliability* 61 (2), pp. 323–335.

Sutharssan, Thamo; Stoyanov, Stoyan; Bailey, Chris; Yin, Chunyan (2015): Prognostic and Health Management for Engineering Systems: A Review of the Data-Driven Approach and Algorithms. In *The Journal of Engineering* 2015 (7), pp. 215–222.

ISO/IEC/IEEE 15288:2015, 2015: Systems and Software Engineering - System Life Cycle Processes.

Taylor, Chris; Bandara, Wasana (2003): Defining the Quality of Business Process Reference Models. In : Proceedings of the Australasian Conference on Information Systems, vol. 14, pp. 1–10.

Thorndike, Robert L. (1953): Who Belongs in the Family? In *Psychometrika* 18 (4), pp. 267–276.

Too Yuen Min (2020): Time Series Distance Measures (ts-dist) - GitHub Repository. Available online at https://github.com/ymtoo/ts-dist, updated on 6/22/2020, checked on 6/22/2020.

Trilla, Alexandre; Dersin, Pierre; Cabré, Xavier (2018): Estimating the Uncertainty of Brake Pad Prognostics for High-Speed Rail with a Neural Network Feature Ensemble. In : Proceedings of the Annual Conference of the PHM Society, vol. 10.

Tsui, Kwok L.; Chen, Nan; Zhou, Qiang; Hai, Yizhen; Wang, Wenbin (2015): Prognostics and Health Management. A Review on Data Driven Approaches. In *Mathematical Problems in Engineering* 2015 (6), pp. 1–17.

Uckun, Serdar; Goebel, Kai; Lucas, Peter J.F. (2008): Standardizing Research Methods for Prognostics. In : Proceedings of International Conference on Prognostics and Health Management. Denver, United States, pp. 1–10, checked on 12/8/2017.

Uluyol, Onder; Parthasarathy, Girija (2012): Multi-Turbine Associative Model for Wind Turbine Performance Monitoring. In : Proceedings of the Annual Conference of the Prognostics and Health Management Society.

Vachtsevanos, George J. (2006): Intelligent Fault Diagnosis and Prognosis for Engineering Systems. Hoboken, United States: Wiley.

Venkatasubramanian, Venkat; Rengaswamy, Raghunathan; Kavuri, Surya N.; Yin, Kewen (2003): A Review of Process Fault Detection and Diagnosis. Part III: Process History Based Methods. In *Computers and Chemical Engineering* 27 (3), pp. 327–346.

Vlachos, Michail; Kollios, George; Gunopulos, Dimitrios (2002): Discovering Similar Multidimensional Trajectories. In : Proceedings of International Conference on Data Engineering, vol. 18. San Jose, United States, pp. 673–684.

Vogl, Gregory W.; Weiss, Brian A.; Donmez, M. A. (2014a): NISTIR 8012. Standards Related to Prognostics and Health Management (PHM) for Manufacturing. National Institut for Standards and Technology.

Vogl, Gregory W.; Weiss, Brian A.; Donmez, M. A. (2014b): Standards for Prognostics and Health Management (PHM) Techniques Within Manufacturing Operations. In : Proceedings of Annual Conference of the Prognostics and Health Management Society. Fort Worth, United States.

Vogl, Gregory W.; Weiss, Brian A.; Helu, Moneer (2016): A Review of Diagnostic and Prognostic Capabilities and Best Practices for Manufacturing. In *Journal of Intelligent Manufacturing* 63 (1), pp. 79–95.

Voisin, Alexandre; Levrat, Eric; Cocheteux, Pierre; Iung, Benoît (2010): Generic Prognosis Model for Proactive Maintenance Decision Support: Application to Pre-Industrial E-Maintenance Test Bed. In *Journal of Intelligent Manufacturing* 21 (2), pp. 177–193.

Voisin, Alexandre; Medina-Oliva, Gabriela; Monnin, Maxime; Leger, Jean-Baptiste; Iung, Benoît (2015): Fault Diagnosis System Based on Ontology for Fleet Case Reused. In Vahid Ebrahimipour, Soumaya Yacout (Eds.): Ontology Modeling in Physical Asset Integrity Management. Cham, Switzerland: Springer, pp. 133–169.

Voisin, Alexandre; Medina-Oliva, Gabriela; Monnin, Maxime; Léger, Jean-Baptiste; Iung, Benoît (2013): Fleet-Wide Diagnostic and Prognostic Assessment. In : Proceedings of Annual Conference of the Prognostics and Health Management Society. New Orleans, United States.

vom Brocke, Jan; Simons, Alexander; Niehaves, Bjoern; Niehaves, Bjorn; Reimer, Kai; Plattfaut, Ralf; Cleven, Anne (2009): Reconstructing the Gian. On the Importance of Rigour in Documenting the Literature Search Process. In *ECIS 2009 Proceedings*.

Walden, David D.; Roedler, Garry J.; Forsberg, Kevin; Hamelin, R. D.; Shortell, Thomas M. (Eds.) (2015): Systems Engineering Handbook. A Guide for System Life Cycle Processes and

Activities. International Council on Systems Engineering. 4th ed. Hoboken, United States: Wiley.

Wang, Pingfeng; Youn, Byeng D.; Hu, Chao (2012): A Generic Probabilistic Framework for Structural Health Prognostics and Uncertainty Management. In *Mechanical Systems and Signal Processing* 28, pp. 622–637.

Wang, Tianyi (2010): Trajectory Similarity Based Prediction for Remaining Useful Life Estimation.

Wang, Tianyi; Jianbo Yu; Siegel, David; Lee, Jay (2008): A Similarity-Based Prognostics Approach for Remaining Useful Life Estimation of Engineered Systems. In : International Conference on Prognostics and Helth Management. Denver, United States, pp. 1–6.

Wang, Xiaoyue; Mueen, Abdullah; Ding, Hui; Trajcevski, Goce; Scheuermann, Peter; Keogh, Eamonn (2013): Experimental Comparison of Representation Methods and Distance Measures for Time Series Data. In *Data Mining and Knowledge Discovery* 26 (2), pp. 275–309.

Wang, Zhao-Qiang; Hu, Chang-Hua; Fan, Hong-Dong (2018): Real-Time Remaining Useful Life Prediction for a Nonlinear Degrading System in Service: Application to Bearing Data. In *IEEE/ASME Transactions on Mechatronics* 23 (1), pp. 211–222.

Webster, Jane; Watson, Richard T. (2002): Analyzing the Past to Prepare for the Future: Writing a Literature Review. In *MIS quarterly*, pp. xiii–xxiii.

Weske, Mathias (2012): Business Process Management. Berlin, Heidelberg, Germany: Springer Berlin Heidelberg.

Widera, Adam; Hellingrath, Bernd (2011): Improving Humanitarian Logistics - Towards a Tool-Based Process Modeling Approach, pp. 273–295.

Winter, Robert; Gericke, Anke; Bucher, Tobias (2009): Method Versus Model - Two Sides of the Same Coin? In Antonia Albani, Joseph Barjis, Jan L.G. Dietz (Eds.): Advances in Enterprise Engineering III, vol. 34. Berlin, Germany: Springer (34), pp. 1–15.

Wirth, Rüdiger; Hipp, Jochen (2000): CRISP-DM. Towards a Standard Process Model for Data Mining. In : Proceedings of International Conference on the Practical Applications of Knowledge Discovery and Data Mining, vol. 4, pp. 29–39.

Wu, Dazhong; Jennings, Connor; Terpenny, Janis; Gao, Robert X.; Kumara, Soundar (2017a): A Comparative Study on Machine Learning Algorithms for Smart Manufacturing: Tool Wear Prediction Using Random Forests. In *Journal of Manufacturing Science and Engineering* 139 (7).

Wu, Yuting; Yuan, Mei; Dong, Shaopeng; Lin, Li; Liu, Yingqi (2017b): Remaining Useful Life Estimation of Engineered Systems Using Vanilla LSTM Neural Networks. In *Neurocomputing* 275.

Xia, Tangbin; Dong, Yifan; Xiao, Lei; Du, Shichang; Pan, Ershun; Xi, Lifeng (2018): Recent Advances in Prognostics and Health Management for Advanced Manufacturing Paradigms. In *Reliability Engineering and System Safety* 178, pp. 255–268.

Xue, Feng; Bonissone, Piero P.; Varma, Anil; Yan, Weizhong; Eklund, Neil; Goebel, Kai (2008): An Instance-Based Method for Remaining Useful Life Estimation for Aircraft Engines. In *Journal of Failure Analysis and Prevention* 8 (2), pp. 199–206.

Yang, Chunsheng; Letourneau, Sylvain; Liu, Jie; Cheng, Qiangqiang; Yang, Yubin (2017): Machine Learning-Based Methods for TTF Estimation with Application to APU Prognostics. In *Applied Intelligence* 46 (1), pp. 227–239.

Ye, Zhi-Sheng; Xie, Min (2015): Stochastic Modelling and Analysis of Degradation for Highly Reliable Products. In *Applied Stochastic Models in Business and Industry* 31 (1), pp. 16–32.

Yu, Peng; Yong, Xu; Datong, Liu; Xiyuan, Peng (2012): Sensor Selection with Grey Correlation Analysis for Remaining Useful Life Evaluation. In : Proceedings of Annual Conference of the Prognostics and Health Management Society. Minneapolis, United States.

Zaidan, Martha A. (2014): Bayesian Approaches for Complex System Prognostics. University of Sheffield, Sheffiled, United Kingdom.

Zaidan, Martha A.; Harrison, Robert F.; Mills, Andrew R.; Fleming, Peter J. (2015a): Bayesian Hierarchical Models for Aerospace Gas Turbine Engine Prognostics. In *Expert Systems with Applications* 42 (1), pp. 539–553.

Zaidan, Martha A.; Mills, Andrew R.; Harrison, Robert F. (2013): Bayesian Framework for Aerospace Gas Turbine Engine Prognostics. In : Proceedings of IEEE Aerospace Conference. Big Sky, United States, pp. 1–8.

Zaidan, Martha A.; Mills, Andrew R.; Harrison, Robert F.; Fleming, Peter J. (2016): Gas Turbine Engine Prognostics Using Bayesian Hierarchical Models: A Variational Approach. In *Mechanical Systems and Signal Processing* 70-71, pp. 120–140.

Zaidan, Martha A.; Relan, Rishi; Mills, Andrew R.; Harrison, Robert F. (2015b): Prognostics of Gas Turbine Engine: An Integrated Approach. In *Expert Systems with Applications* 42 (22), pp. 8472–8483.

Zhang, Guigang; Wang, Jian; Lv, Zhi; Yang, Yi; Su, Haixia; Yao, Qi et al. (2015): A Integrated Vehicle Health Management Framework for Aircraft. A Preliminary Report. In : Proceedings of IEEE Conference on Prognostics and Health Management. Austin, United States, pp. 1–8.

Zhang, Jingliang; Lee, Jay (2011): A Review on Prognostics and Health Monitoring of Li-ion Battery. In *Journal of Power Sources* 196 (15), pp. 6007–6014.

Zhao, Rui; Yan, Ruqiang; Wang, Jinjiang; Mao, Kezhi (2017): Learning to Monitor Machine Health with Convolutional Bi-Directional LSTM Networks. In *Sensors* 17 (2).

Zhou, Yingya; Chioua, Moncef; Ni, Weidou (2016): Data-Driven Multi-Unit Monitoring Scheme With Hierarchical Fault Detection and Diagnosis. In : Proceedings of Mediterranean Conference on Control and Automation, vol. 24. Athens, Greece, pp. 13–18.

Zio, Enrico; Di Maio, Francesco (2010): A Data-Driven Fuzzy Approach for Predicting the Remaining Useful Life in Dynamic Failure Scenarios of a Nuclear System. In *Reliability Engineering & System Safety* 95 (1), pp. 49–57.

Zorn, Ted; Campbell, Nittaya (2006): Improving The Writing Of Literature Reviews Through A Literature Integration Exercise. In *Business Communication Quarterly* 69 (2), pp. 172–183.

Zuccolotto, Marcos; Pereira, Carlos E.; Fasanotti, Luca; Cavalieri, Sergio; Lee, Jay (2015): Designing an Artificial Immune Systems for Intelligent Maintenance Systems. In *IFAC-PapersOnLine* 48 (3), pp. 1451–1456.

Appendix

A   Literature Search Process

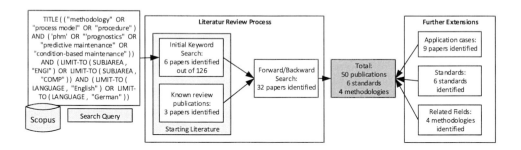

Figure 72 | Literature Search Process RQ1

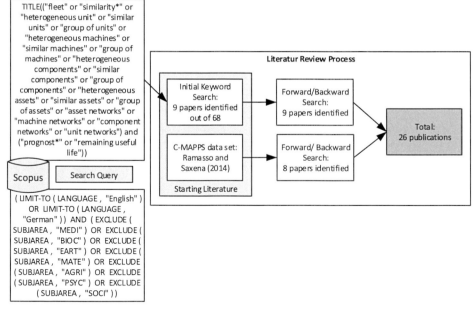

Figure 73 | Literature Search Process RQ2

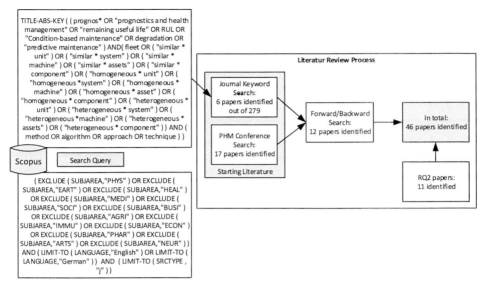

Figure 74 | Literature Search Process RQ3

B   In- and Outputs of the Process Reference Model

Table 31 | Required Inputs and Derived Outputs on the Process Level

Phase	Processes	Required Inputs	Derived Outputs
Preparation Phase	Project Specification		Target Definition, Stakeholder Requirements, Project Charter, Project Specification & System Constraints, Detailed (System) Requirements, Project Plan
Preparation Phase	As-Is Modelling		As-Is Maintenance Processes
Preparation Phase	Candidate Identification		Equipment List PdM Candidate List, Prioritized PdM Candidate List
Preparation Phase	Prerequisite Determination		Failure Logic & Failure Modes, Parameter Measurement
Preparation Phase	Measurement Extension		Measurement Techniques, Measurement Location
PHM Design Phase	Concept Elaboration	Detailed Requirements	Candidate Requirements, Performance Metrics, Candidate Concept, Use Case Description
PHM Design Phase	Engineering Knowledge Formulation		Normal Behavior Description, Causes/ Consequences, Candidate Description
PHM Design Phase	Data Preprocessing	Detailed Requirements, Parameter Measurements	Data Documentation
PHM Design Phase	Algorithm Development	Candidate Description	Model Assumption, Test Design, PHM Algorithm(s) incl. Parameters
PHM Design Phase	Algorithm Assessment	Performance Metrics, Candidate Requirements, Detailed Requirements, Project Specification & System Constraints	
PHM Design Phase	Solution Mitigation	Candidate Requirements	

System Architecture Design Phase	Objectives and Requirement Analysis	Project Specification & System Constraints, Detailed Requirements	System Specification, System Architecture Requirements
	Physical Architecture Design	System Architecture Requirements	Physical Architecture Specification
	Logical Architecture Design	System Architecture Requirements	Logical Architecture Specification
	System Implementation	System Architecture Requirements	
Deployment Phase	Deployment Preparation		Deployment Plan
	Process Adjustments	As-Is Maintenance Processes	To-Be Maintenance Processes, Process Organization
	Candidate Solution Deployment	Deployment Plan	
	Sustainment Implementation		Monitoring and Maintenance Process Plan
	Candidate Completion		Candidate Documentation, Experience Documentation
Project Completion	Completion Preparation		Final Presentation, Final Report, Documentation
	Project Review	Experience Documentation	Experience Documentation

C   Reference View of Process Reference Model

The reference view of the process reference model shows the identified process elements grouped with regard to their processes and phases. In addition, it highlights the main outputs on the process level, as well as the project management tasks.

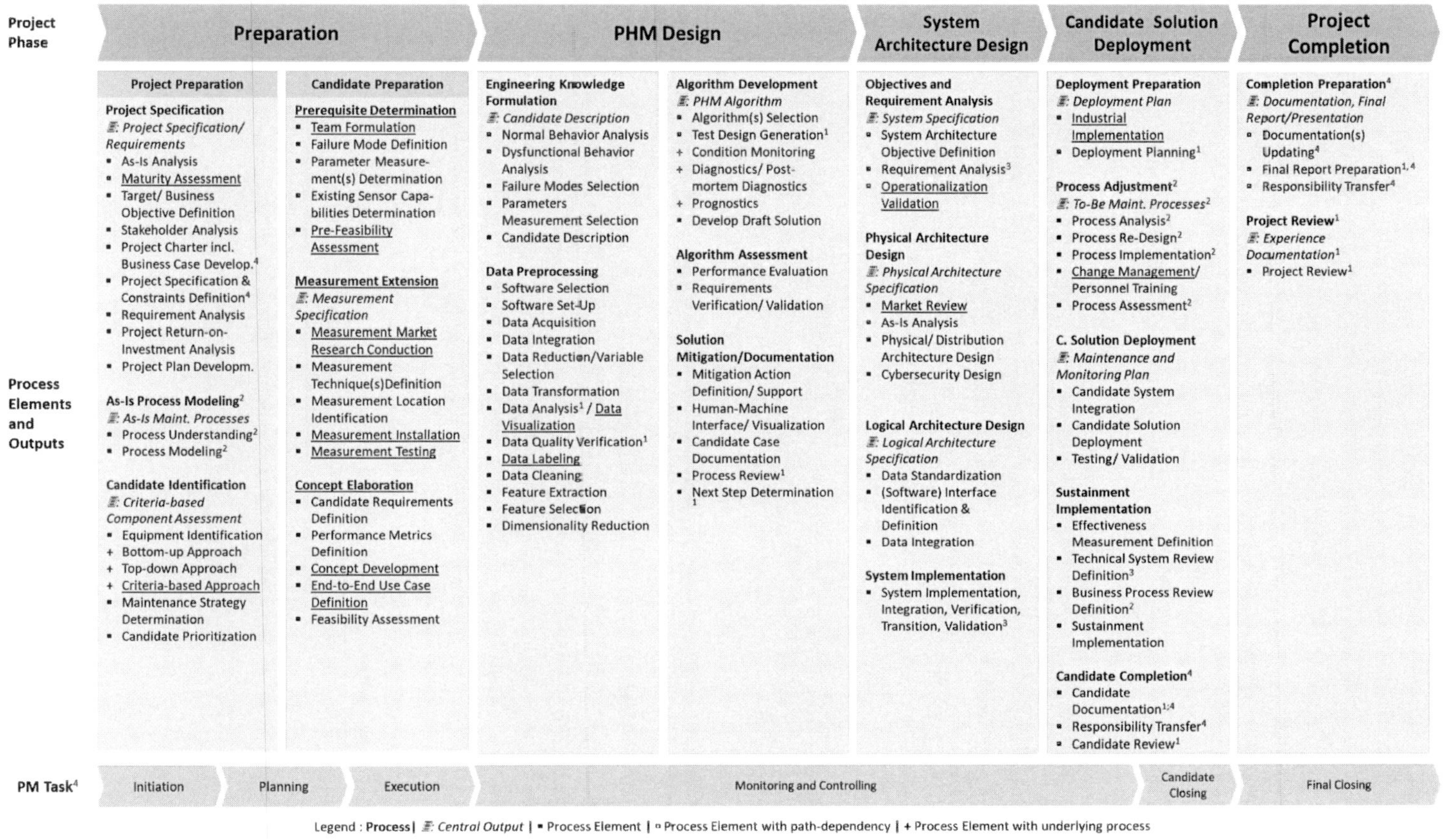

Project Phase

Preparation	PHM Design	System Architecture Design	Candidate Solution Deployment	Project Completion

Process Elements and Outputs

Preparation

Project Preparation
- *Project Specification/ Requirements*
  - Project Specification[2]
  - As-Is Analysis
  - Maturity Assessment
  - Target/ Business Objective Definition
  - Stakeholder Analysis
  - Project Charter incl. Business Case Develop.[4]
  - Project Specification & Constraints Definition[4]
  - Requirement Analysis
  - Project Return-on-Investment Analysis
  - Project Plan Developm.

As-Is Process Modeling[2]
- *As-Is Maint. Processes*
  - Process Understanding[2]
  - Process Modeling[2]

Candidate Identification
- *Criteria-based Component Assessment*
  + Equipment Identification
  + Bottom-up Approach
  + Top-down Approach
  + Criteria-based Approach
  - Maintenance Strategy Determination
  - Candidate Prioritization

Candidate Preparation
- **Prerequisite Determination**
  - Team Formulation
  - Failure Mode Definition
  - Parameter Measurement(s) Determination
  - Existing Sensor Capabilities Determination
  - Pre-Feasibility Assessment

Measurement Extension
- *Measurement Specification*
  - Measurement Market Research Conduction
  - Measurement Technique(s) Definition
  - Measurement Location Identification
  - Measurement Installation
  - Measurement Testing

Concept Elaboration
  - Candidate Requirements Definition
  - Performance Metrics Definition
  - Concept Development
  - End-to-End Use Case Definition
  - Feasibility Assessment

PHM Design

Engineering Knowledge Formulation
- *Candidate Description*
  - Normal Behavior Analysis
  - Dysfunctional Behavior Analysis
  - Failure Modes Selection
  - Parameters Measurement Selection
  - Candidate Description

Data Preprocessing
  - Software Selection
  - Software Set-Up
  - Data Acquisition
  - Data Integration
  - Data Reduction/Variable Selection
  - Data Transformation
  - Data Analysis / Data Visualization
  - Data Quality Verification[1]
  - Data Labeling
  - Data Cleaning
  - Feature Extraction
  - Feature Selection
  - Dimensionality Reduction

Algorithm Development
- *PHM Algorithm*
  - Algorithm(s) Selection[1]
  - Test Design Generation[1]
  + Condition Monitoring
  - Diagnostics/ Post-mortem Diagnostics
  + Prognostics
  - Develop Draft Solution

Algorithm Assessment
  - Performance Evaluation
  - Requirements Verification/Validation

Solution Mitigation/Documentation
  - Mitigation Action Definition/ Support
  - Human-Machine Interface/ Visualization
  - Candidate Case Documentation
  - Process Review[1]
  - Next Step Determination[1]

System Architecture Design

Objectives and Requirement Analysis
- *System Specification*
  - System Architecture Objective Definition
  - Requirement Analysis[3]
  - Operationalization Validation

Physical Architecture Design
- *Physical Architecture Specification*
  - Market Review
  - As-Is Analysis
  - Physical/ Distribution Architecture Design
  - Cybersecurity Design

Logical Architecture Design
- *Logical Architecture Specification*
  - Data Standardization
  - (Software) Interface Identification & Definition
  - Data Integration

System Implementation
  - System Implementation, Integration, Verification, Transition, Validation[3]

Candidate Solution Deployment

Deployment Preparation
- *Deployment Plan*
  - Industrial Implementation
  - Deployment Planning[1]

Process Adjustment[2]
- *To-Be Maint. Processes[2]*
  - Process Analysis[2]
  - Process Re-Design[2]
  - Process Implementation[2]
  - Change Management/ Personnel Training
  - Process Assessment[2]

C. Solution Deployment
- *Maintenance and Monitoring Plan*
  - Candidate System Integration
  - Candidate Solution Deployment
  - Testing/ Validation

Sustainment Implementation
  - Effectiveness Measurement Definition
  - Technical System Review Definition[3]
  - Business Process Review Definition[2]
  - Sustainment Implementation

Candidate Completion[4]
  - Candidate Documentation[1,4]
  - Responsibility Transfer[4]
  - Candidate Review[1]

Project Completion

Completion Preparation[4]
- *Documentation, Final Report/Presentation*
  - Documentation(s) Updating[4]
  - Final Report Preparation[1,4]
  - Responsibility Transfer[4]

Project Review[1]
- *Experience Documentation[1]*
  - Project Review[1]

PM Task[4]

Initiation	Planning	Execution	Monitoring and Controlling	Candidate Closing	Final Closing

Legend : ▪ Process | ≣ Central Output | ▪ Process Element | ° Process Element with path-dependency | + Process Element with underlying process

Figure 75 | Reference View of Process Reference Model

D   Detailed Process of Reference Model

In this section, the detailed elaborated processes on level 3 are provided. For documentation requirements, phases, processes, and process elements, which have been added by the inclusion of the methodologies of the related fields, are marked with a number, referring to the respective methodology (1 – Crisp-DM, 2 – MIPI, 3 – INCOSE, 4 – PMBOK). Furthermore, adjustments made as a result of the evaluation are underlined in the reference view and written in italics in the detailed models.

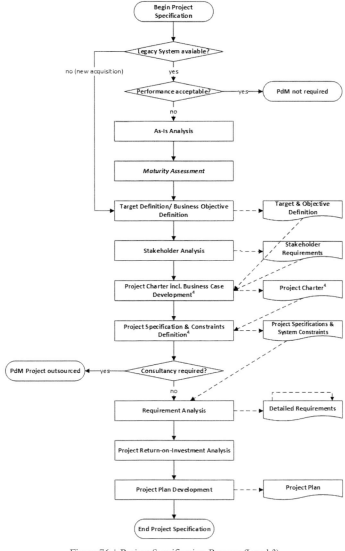

Figure 76 | Project Specification Process (Level 3)

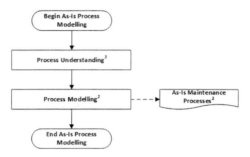

Figure 77 | As-Is Process Modeling (Level 3)

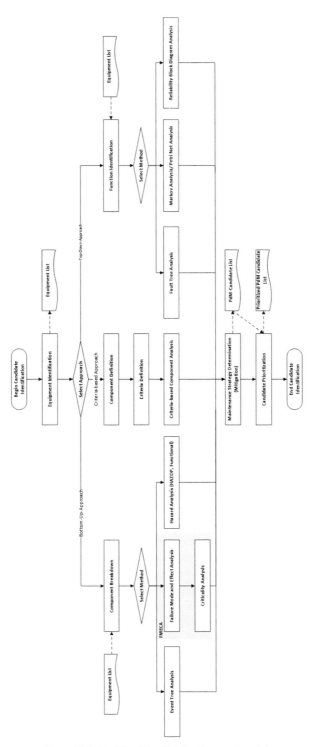

Figure 78 | Candidate Identification Process (Level 3)

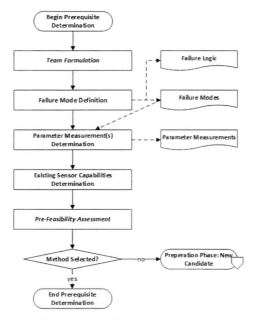

Figure 79 | Prerequisite Determination Process (Level 3)

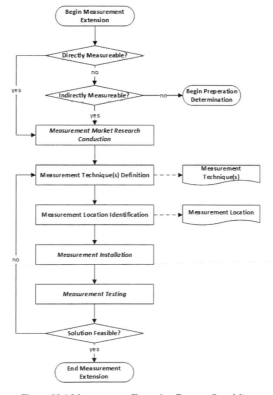

Figure 80 | Measurement Extension Process (Level 3)

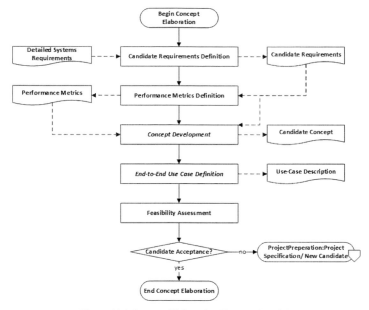

Figure 81 | Concept Elaboration Process (Level 3)

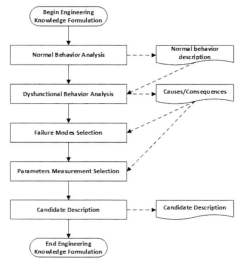

Figure 82 | Engineering Knowledge Formulation Process (Level 3)

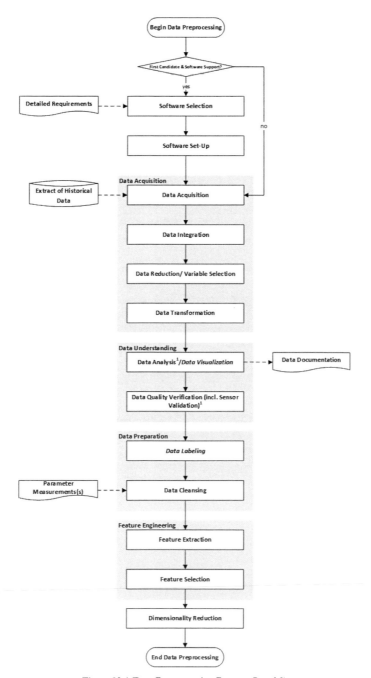

Figure 83 | Data Preprocessing Process (Level 3)

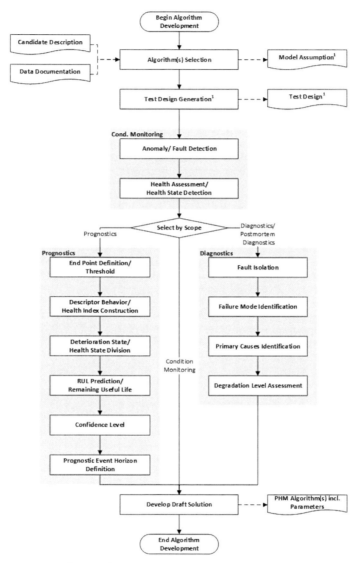

Figure 84 | Algorithm Development Process (Level 3)

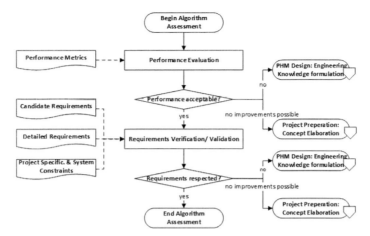

Figure 85 | Algorithm Assessment Process (Level 3)

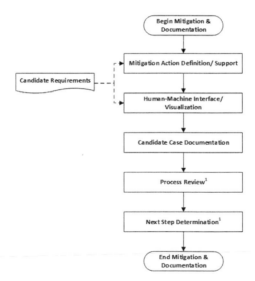

Figure 86 | Mitigation and Documentation Process (Level 3)

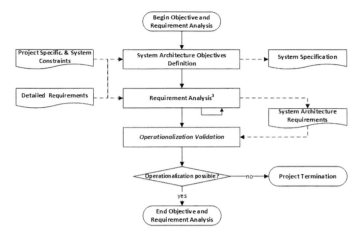

Figure 87 | Objectives and Requirements Analysis Process (Level 3)

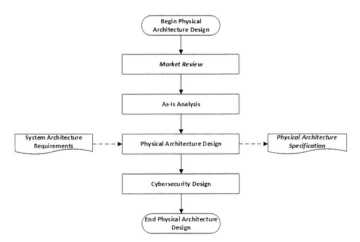

Figure 88 | Physical Architecture Design Process (Level 3)

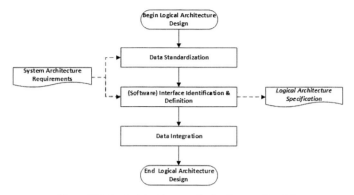

Figure 89 | Logical Architecture Design Process (Level 3)

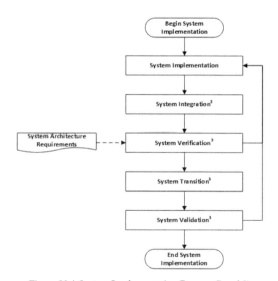

Figure 90 | System Implementation Process (Level 3)

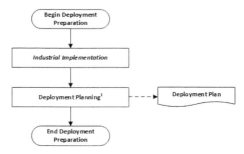

Figure 91 | Deployment Preparation Process (Level 3)

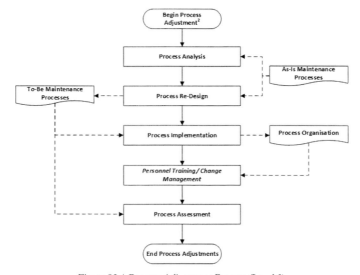

Figure 92 | Process Adjustment Process (Level 3)

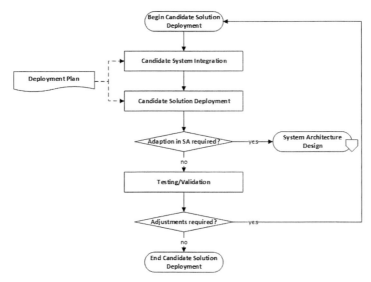

Figure 93 | Candidate Solution Deployment Process (Level 3)

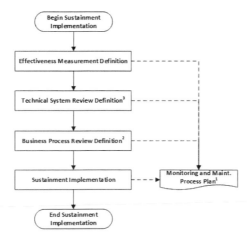

Figure 94 | Sustainment Implementation Process (Level 3)

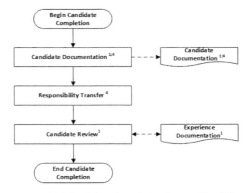

Figure 95 | Candidate Completion Process (Level 3)

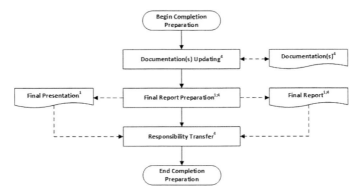

Figure 96 | Completion Preparation Process (Level 3)

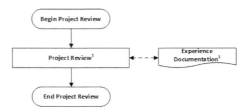

Figure 97 | Project Review Process (Level 3)

E   Expert Interview Guideline

Table 32 | Expert Interview Guideline

Phase	Topic	Main Questions
Warm-up phase	-	Could you please shortly introduce yourself, the company you are working for and your position in the company? Could you shortly explain your general tasks and responsibilities? How is your work connected to PdM, and how much experience do you have in this domain?
Main phase	PM introduction process	Could you explain your PdM implementation process? Which distinct phases can be identified within this process?
Main phase	Project Preparation	Who is the initiator of a PM project? How is the identification of objects for PdM realized? Is there a business case and a formal definition of business and technical requirements? Why not? Which steps are included? What are the different roles that are required or have been determined in a PdM introduction project?
Main phase	PHM Design	How is the appropriate algorithm to utilize the data chosen or designed? Based on which techniques (e.g., clustering) does the PM solution diagnose or predict failures? Which techniques have been tested?
Main phase	System Architecture Design	How are the relevant data sources identified? How is the data stored (e g., one database)? Which steps are necessary in order to prepare and harmonize the data?
Main phase	Candidate Solution Deployment/ Application	How does the new PM system influence the existing maintenance process? In which way does the existing process have to be adapted regarding its existing tasks, roles, and artifacts (e.g., application systems)? Are there any training measures planned for employees? Are there any planned steps after the rollout of the PdM solution, such as Continuous Improvement?
Main phase	[All]	Potential follow-up questions: Which particular outcome has this phase? To what extent are these phases/ steps a pre-defined process by your company or based on the project participant's intuition and experience? Are there dependencies or loops between phases? Are there any tools or models applied to support or structure the overall PdM introduction process or single steps (e.g., CRISP-DM)? To which extent are the project proceedings and results documented for re-use in subsequent PM projects? What is planned for the future? Challenges in each stage?

F   Expert Interview Descriptions

Company A: Steel Tube Manufacturer

Process Description: Company A is in the early stage of its predictive maintenance implementation. The project was initiated two years before the interview by the management with the aim to stay competitive and remain at the cutting edge of technology. Its main target is the reliability improvement of different production machines. In a preliminary step, the exploration of the resulting benefits for the company was paramount. These were refined during requirements analysis by the definition of expectations in terms of improved machine reliability and process improvements. Based on historical failure rates and process criticality, machines are selected for

PdM consideration. Even though there is already a variety of data available, further market research was conducted on additional measurement techniques. Selected sensors are installed and tested. The company already records data for the regular operation of the machine. For the PdM project, data visualization software was therefore introduced to facilitate data analysis by providing a graphical representation of the data over time. In the first step, the company analyzed unplanned machine failures in order to understand the root causes and find measures to prevent failure in the future. The analysis is performed manually through data inspection; however, data-driven algorithms are envisioned for future applications. Based on these results, reports will be generated and send directly to the responsible persons. In addition, the transfer of the solution to the production process is planned via personnel training. The company already implements a continuous review process for process optimization, which will be responsible for the PdM solution.

The process is dynamic and allows for return flows. The project team consists of employees from the technical, production, and welding technology department. In addition, support by the management, maintenance personnel, and machine operators is required.

Assessment and Adaptations: The described process is well-aligned with the developed process reference model. Due to the low maturity of the company with regard to PdM, only a fraction of the process is, however, implemented. As the company has already established a good machine and data connectivity prior to the project, the system architecture design phase is not targeted. In addition, PHM design is only considered in terms of manual post-mortem diagnostics, while the deployment phase is described in a visionary view. Changes to the process reference model resulting from the interview are the measurement market research conduct with sensor installation and testing (added to the measurement extension process) as well as the manual data analysis/ data visualization (refined in the data preprocessing process). While the selection and set-up of the software support might be performed prior to the PHM design phase, it is introduced in the process reference model only at the point when it becomes essential.

Company B: Automotive Manufacturer

Process Description: The PdM project of Company B addresses the different machinery in their production process. The described process depicts the approach followed by one plant, which is, however, independent from their research center. Nevertheless, several steps highlight cooperation and information exchange between different plants and the research center. In an initial step, an as-is analysis is performed, which targets the identification of existing cooperations, the state of technology, involved service providers, available measurements and failure documentations, as well as possible restrictions and constraints. Based on these results, requirements are defined, and a machine is selected for an initial proof-of-concept. The selection criteria include both the machine criticality and failure rate as well as the technical data accessibility. By communication with other plants, information with regard to sensor technology, hardware, and concepts are exchanged and aligned. For this reason, no separate cost-benefit analysis was carried out. This information supported the development of a concept for the basic system architecture. Subsequently, fault causes are identified, and market analysis is conducted to determine adequate sensors and hardware components. For the proof-of-concept, vibration and temperature sensors are installed, and a technical infrastructure via edge components and a central database are established. The collected data was pre-processed afterward and transferred into a usable format. To provide decision support for the maintenance department, data is graphically visualized as a time-series, and a simple dashboard is designed with an automatic notification feature in case of outliers. In the future, further development of diagnostic and prognostic methods is envisioned using Machine Learning techniques. For this reason, a baseline should be defined with associated warn and stop limits. In addition, it is considered to outsource the project. However, a basic understanding of the topic should be available to better assess potential service providers and to be capable of clarifying legal aspects.

The described process has been executed in a very iterative manner. Due to the cooperation of different departments and their restrictions, it was necessary to revise previous decisions several times. The project is led by the technical planning department in close cooperation with the maintenance department. Furthermore, coordination with the IT department, in particular the database management and network management, was required.

Assessment and Adaptations: The process of Company B is in accordance with the developed process reference model. Similar to Company A, the stated process shows a low degree of maturity with only data visualization and condition monitoring currently implemented in the PHM design phase. At the time of the interview, the company was in the PHM design and system architecture design phase; for this reason, the deployment phase has not been targeted yet. Deviations from the process reference model are the concrete designation of the concept development step, which was added after the requirements. While, in the case of Company B, the concept does not target specific fault causes, which are only identified afterward, in the general PdM reference process, possible fault modes are specified prior to an improved design and elaboration of the concept. In addition to the concept development step, the review of the hardware market was highlighted by Company B and supplemented to the reference process in the physical architecture design process. Lastly, the data visualization step corresponds to the data analysis step of the process reference model.

Company C: Aircraft Engine Manufacturer

Process Description: Company C develops an approach to improve the maintenance of its manufacturing assets. Instead of considering entire machines, Company C initially identifies and groups its production assets with regard to different assembly groups. Machines within the same assembly group exhibit high similarity in their structure and functionality. Based on an evaluation matrix consisting of six production- and asset-related criteria, the identified assembly groups, are classified into three different service levels. Each level defines its maintenance requirements, with predictive maintenance being the first service level. Beyond that, the systematic optimization process might reveal problematic machines in terms of high failure frequency or criticality, which are additionally considered for predictive maintenance. With the support of experts, four candidates are identified for a proof-of-concept. While cost-benefit analysis is envisioned for potential PdM machines, it has not been conducted as currently, no justification is necessary. For each candidate, crucial parameters for failure detection are identified with the help of engineers and electronic technicians. In case of insufficient data, additional sensors are equipped. In parallel, an internal data standard for machinery is developed, including sensor requirements. Relevant identified data is pre-processed using basic operations. The company stores data locally at the machine for three months. A lower data resolution is transferred afterward to a database. To test various application systems, different software providers are considered for each pilot candidate, which allows for visualization and provides analytical capabilities. At the time of the interview, the digital transformation manager developed the first models for anomaly detection and diagnostics. However, due to a lack of knowledge, a company-wide competence center for data analytics will be established in the future. To evaluate the performance and improve acceptance, at first, the results of the implemented model are reviewed manually in the data and at the respective machine. Possible deficiencies are redirected to the model development step, and performed maintenance activities are documented. A dashboard with different user views and possible maintenance instructions is envisioned for the deployed models. This goes along with required process adjustments. Instead of reacting to machine failures, experts monitor the data and receive automated warnings prior to asset failure. Maintenance personnel is instructed thereupon with the repair operations to be carried out. For a successful implementation of PdM, Company C emphasizes the need for change management. To best benefit from the project, the mindset of maintenance technicians has to change, and the willingness to adapt to new technologies is crucial. In regards to the project, working documentation and management summaries are realized.

The project follows a pre-defined process with several possible return flows, which were developed based on experiences from previous projects and in discussion with other companies. Initially, four pilot candidates are considered; however, other assembly groups and machines are targeted thereafter. The project is carried out by the companies own project department in an interdisciplinary team, including the asset manager, electronics technicians, the digital transformation manager, and the maintenance department. To coordinate the project, the four-phase project management model is applied.

Assessment and Adaptations: The process of Company C largely corresponds to the theoretical process reference model. All phases are covered but to a different extent. Similar to Company A and B, company C has a low degree of maturity with regard to the PHM design phase. In addition, the system architecture phase has not been targeted in detail; however, adaptations of the current structure are considered. In contrast to the other processes, the company approaches the project from a higher level. Besides the identification of PdM candidates, appropriate maintenance strategies are defined for all equipment. Furthermore, assembly groups are identified, which refer to components of the process reference model. To emphasize the definition of PdM components for the criteria-based approach in the equipment and function sub-process, the step component definition was added. Another difference depicts the manual performance evaluation step. While performance evaluation with regard to performance metrics is included, the step can also be realized differently. Beyond that, the application phase covers a feedback loop to model development in case of deviations. As part of a test phase, the manual review of the results can be implemented. Lastly, the step change management is added to the deployment preparation process.

Company D: Aircraft Manufacturer

Process Description: The PdM project of company D aims to provide services to its customers via a company-wide digital platform. Depending on the purchased service package, the customers are offered different functionalities to monitor the health status of aircraft components. The following process depicts the project for cabin elements, which is, however, aligned with the general company process. In the first step, possible candidates are identified. In the case of the cabin, these are replaceable line components that are exchangeable and non-integral parts of the aircraft. Based on reliability statistics and costs per flight hours, potential candidates are identified and prioritized by the customer service. Beyond that, candidates with high data availability are determined. With regard to a specific candidate, a team consisting of the customer service, engineering, design office, and data scientists is defined. Thereafter, possible failure modes and their failure logic are identified, which is followed by a pre-feasibility analysis. This analysis includes analytical feasibility, in terms of the possibility to identify and predict different failure modes by an analytical model, as well as historical data availability. In case the candidate proves to be feasible and relevant, an initial prototype is developed based on the failure logic. At this stage, these approaches are rather heuristic. In addition, the formal specifications of the model are documented. The developed model is tested for a specific time interval, and its performance is assessed based on pre-defined measures. If the performance requirements are not achieved, model improvements are implemented. Completing the model development, the final decision for continuation is to be made by the decision-making committee. With the support of professional programmers, industrial implementation of the solution is programmed based on the specified requirements. The candidate solution is then integrated into the platform and provided to their customers. Models are monitored post-integration continuously. However, no systematic feedback channel is available.

The process allows for several return loops, in particular between the prototype development and its validation. For each candidate, a separate project team is defined, including members from customer service, the program management of the aircraft model, the engineering department, data analysts, and software programmers. In addition, there is close cooperation with external service providers for implementation and data analysis. The project is managed in a flexible and agile manner with guiding values provided by the company. Kanban-Boards are used to monitor project status.

Assessment and Adaptations: The described process provides a structured approach that responds well to existing needs. The process is designed on the company level and is therefore well-established and proven by different applications. Nonetheless, it is largely consistent with the theoretical process reference model. Due to a lack of data analytic competencies, developed models currently show a rather low degree of maturity and rely on identified failure logic only. In addition, the existing data infrastructure provides for data to be transferred as soon as the aircraft reaches ground level. Only safety-critical data is transmitted during the flight. For this reason, the system architecture design is not targeted. Nevertheless, some modifications of the reference model result from the interview. Due to the heterogeneous candidates, different project teams are formulated. This step should be performed prior to measurement specifications, however, with relation to a specific candidate. Another adjustment depicts the pre-feasibility assessment. Both steps are inserted in the prerequisite determination phase. Finally, the industrial implementation was supplemented as a step in the deployment preparation.

Company E: Wind Turbine Manufacturer

Process Description: Company E develops a large number of PdM solutions for various components of their wind turbines. These models are offered to their customers as additional services as part of existing maintenance contracts. After initial difficulties, a structured process for the implementation of the project was developed and proved to be successful. In the first step of this process, existing problems are identified based on statistical cost assessments from turbine downtimes and repairs. In addition, further possible candidates are selected based on opportunistic possibilities driven either by the customer or employees. Candidates are prioritized and successively realized. For each candidate, a team is designated, which combines domain expertise with data analytic capabilities. After an initial feasibility assessment and data preparation step to clean data from specific operating and environmental conditions, possible symptoms that have preceded certain failures in the past are identified with the help of domain expertise. This results in a precise problem and symptom description. In case no symptom is visible in the data, the candidate is neglected as sensor extensions are too expensive and therefore only aimed at during turbine development. Thereafter, a prediction model is built using machine learning techniques, and detailed documentation of the case is prepared. In addition, resulting maintenance instructions are defined, which require approval from the engineering department. Once several models are completed, they are released for field testing. For this purpose, maintenance technicians of a specific region are trained via online calls, and documentation is provided. Due to infrequent failure occurrence, the test region includes a large number of wind turbines. The technicians contribute very valuable feedback, which can lead to model or documentation adjustments. As soon as the confidence in the model is sufficiently high, it is rolled out on a global level in combination with company-wide employee training. In case symptoms are detected, technicians are informed via a ticketing system. The solutions are monitored continuously, and feedback is received as a comment in the corresponding tickets.

The company emphasized the importance of a standardized process for its PdM project. While the project was initially largely rejected, it was well-received after the structured process has been developed. The process allows for various return flows, in particular possible modifications of the model resulting from feedback in the field testing and roll-out steps. The project is managed and initiated by the data analytics department. The specific candidate teams additionally include experts from the engineering department as well as from field service.

Adaptations: The process of Company E is very mature and reflects well the developed process reference model. All phases are covered except the systems architecture phase, as the company already has an adequate architecture implemented. While several steps are named slightly differently, the main distinction can be seen with regard to employee training and following field testing. However, these steps are very specific to the context of the project and can therefore be subsumed by the process reference model with regard to the test and validation step within the deployment process. Analogous to the previous process from Company D, the steps of team formulation and pre-feasibility analysis are included in the process reference model.

Company F: Production Engineering Research Institute

Process Description: Company F is a research institute, which provides consultancy services to manufacturing companies. They have assisted several companies with their PdM projects, with a particular focus on data analytics. In order to structure the procedure, their approach is based on the CRISP-DM process. The first step within a PdM project is, therefore, the use case definition. In most cases, the use case is already identified by the customer; however, a refinement is necessary in terms of scoping the solution (analytical level definition). The selection of the machinery is often ad-hoc, without a detailed business and cost analysis and mainly based on the highest possible coverage (scale effects). Thereafter, data is assessed. Herefore, available sensors are identified and checked against the use case. This is done in cooperation with domain experts in order to understand the data and its relationships between different sources. The relevant data is acquired, and the storage location for data access is determined. Only then, the data preparation is conducted. This often constitutes the main effort and might lead to project abortion. As an outcome of this process, a formalized description of the data, its characteristics, and aggregation is provided. As soon as data is sufficiently prepared, a variety of different methods are applied in the data mining step, depending on the specified use case, resulting in a PHM model. Methods might range from classification techniques with regard to pattern recognition and machine learning to time-series regression with ARMA. Often multiple models are implemented using a standard toolbox. Fine-tuning has been performed manually afterward. In case fundamental data understanding is lacking, identification of similar machines using clustering is conducted as a preliminary step. The final model is assessed and evaluated. If the quality is insufficient, based on the initial problem definition with regard to different performance metrics, either model improvements are pursued, or a different method is used. Lastly, the project is documented both externally for the company and internally for feedback and knowledge transfer.

The process is described as cyclic and highly iterative. Several initial ideas are discarded, and iterations are required to identify the problem. Furthermore, returns are required if data is too limited or erroneous or the selected model is not able to reach the defined target. The core project team includes data analytics researchers. They are supported by qualified domain experts from the customers (machine designers and operators), as well as internal IT representatives. In addition, management support and decision making were described as highly important for the successful realization. As a project management tool, agile methods are chosen (mainly SCRUM and Kanban).

Assessment and Adaptations: The process is well-aligned with the CRISP-DM methodology and thus provides a rather mature approach. As the company provides assistance for data analytics projects, mainly the PHM design phase is targeted. Due to the consistency with CRISP-DM, the process is well-aligned with the process reference model. One exception depicts the data assessment, which is conducted before the measurement extension. This step has been summarized by the pre-feasibility analysis, which was also described by a few other companies. Furthermore, the data description is added as an output of the data analysis step.

Company G: Management Consultancy

Process Description: Company G is a management consultancy. Consultancy services within PdM are positioned in their technology department in the area of Industrie 4.0 and digital operations analytics. A PdM project at the customer is initiated with as-is business analysis, in which the customer's current state is evaluated. Based on this information, possible use cases are analyzed and assessed. A suitable case is selected afterward for a pilot phase, which provides the highest benefit for the company. In addition, a business case is developed. The pilot phase is structured using the CRISP-DM methodology. At first, the problem is formulated in terms of the technical objective, and requirements are deduced from this. These are evaluated after the pilot phase. This is followed by data preparation, which includes data screening and data labeling. Here, an extract of historical data is provided to the management consultancy team. Thereafter, model development is conducted in close cooperation with the customer in order to integrate future users. The resulting model is evaluated with expert knowledge from the customer and matched against the customer's requirements. These steps are performed iteratively, usually within one sprint of 2-3 weeks duration, until the model performance is sufficient. In parallel to the model development, validation of operationalization is checked, including, among other real-time data access and required infrastructure. This is done under the usage of the waterfall methodology. After the model development and operationalization validation has terminated, a decision for continuation is reached. If continuation is not deemed useful, the project is aborted. In the other case, further use cases are targeted. In addition, a cost-benefit analysis is conducted. Here, benefits are assesses based on the performance of the developed model. The final solution includes a decision-support for the maintenance personnel in terms of visualization of the machinery's health, while the decision is still within the responsibility of the user. This is due to the fact that the existing technical infrastructure is often insufficient for PdM, and an operationalization concept is developed only after the successful completion of the pilot phase. This usually includes a cloud solution in which the models run on the edge component. After deploying the model, continuous model enhancement and re-training are envisioned.

The process is very iterative and dynamic with different adjustment possibilities. It is structured with regard to the two methodologies, CRISP-DM and waterfall model, using sprints every 2-3 weeks. The core project team consists of employees of the management consultancy with expertise and data science, engineering, and information technology (in particular IoT and Cloud technologies). This team is supported during the PHM design phase by the company's maintenance department, the machine operator, and the quality control, as well as by the IT and Production-IT during the operationalization.

Assessment and Adaptations: The process depicts a well-established and mature approach to PdM realizations. It covers all phases, with the exception of the project completion, which was explicitly mentioned. It is in accordance with the process reference model and therefore resulted only in minor adjustments, which were added to the process reference model. This includes the missing steps data labeling (added to the data preprocessing process), operationalization validation, and the subsequent decision (added to the system architecture design), as well as cost-benefit analysis, which has only been introduced in the process reference model during the project preparation. The cost-benefit analysis, however, was not moved towards the candidate solution phase, as this denotes a review of the existing cost-benefit analysis based on actual values given the model performance.

Company H: Management Consultancy

Process Description: Company H is another management consultancy, which provides support and takes over PdM projects for its customers. They are offering managed services, with only an interface provided to the customer. The system is completely managed and maintained by the consultancy. Nevertheless, no own software solution is available, but existing software is used. The project is structured in three phases. In the initialization phase, the design of PdM is demonstrated in one exemplary use case. This phase, therefore, starts with an as-is analysis, where different machine tasks and issues are identified. This is followed by the selection of a machine and requirements definition. Then, if desired by the customer, a cost-benefit analysis can be conducted; however, only on historical values. This is followed by data sighting, which includes identification of sensors, the definition of suitable data, and exploration of these data and data preparation. Sensor extensions are possible but usually not within the scope due to difficulties for re-certification of machines. The available database is used subsequently for anomaly detection and fault cluster identification (pattern). For model development, either machine learning or statistical methods are used. Different models are evaluated with regard to customer feedback and cross-validation. In addition, requirements are matched for model validation. Again, a cost-benefit analysis is possible to highlight the improved benefit via PdM. The phase is completed with a recommendation on the usage. If the recommendation is accepted by the customer, the project is continued with additional use cases and the implementation phase. In the implementation phase, the technical infrastructure is addressed. Data is collected in a time-series database and pre-processed. Data can be access continuously via a data interface. In addition, a dashboard is designed for the operators. To improve usability and acceptance, the final user is included in this process. This usually includes data visualization and operator comments of similar faults. Lastly, in the launch phase, the solution is deployed, and continuous improvement measurements are implemented, which covers model re-evaluation and adaptation.

The process is dynamic with short feedback loops, which allows for returns. Most interactions are required during the pilot phase. Besides the project team of company H, the support team consists of a contact person in the production, the machine operator, the maintenance department as well as the management to release the budget and for project support.

Assessment and Adaptations: The described process for Company H presents a high degree of maturity and is well-established. It is in accordance with the process reference model, however, in a more sequential manner, with no parallel execution of the system architecture phase. Minor adjustments were performed to the process reference model under consideration of the process of Company H. This includes only the operationalization validation (identical to Company G), which has been added to the process reference model.

Company I: Management Consultancy

Process Description: Company I is a management consultancy. Services in the field of PdM are placed in the area of Digitization and Supply Chain and Operations Management. They are offering comprehensive solutions, which are usually targeting original equipment manufacturers. PdM projects are usually initialized by the production department (chief operating officer) or by the service department (asset or service manager). In order to scope the project, primary requirements analysis is performed, and a PdM target is formulated. In particular, the business value drivers and the PdM target are defined. Thereafter, the available data, in terms of the suitability in the context of PdM, and IT infrastructure, with regard to processing and storage of high-frequency data, is assessed. Furthermore, it is investigated whether the installation of further sensors is required. This is followed by the development of a functional design. In this step, the scope of the PdM solution is refined (e.g., visual indicator or fully automated and integrated solution), and the benefit is depicted. In addition, the impact PdM will create for the company, and the entire value chain is highlighted using an end-to-end use case description. Operational and managerial implications are depicted, and responsibilities are defined. Using this information, subsequently, the maturity is assessed, gathering additional data from the customer on the current maintenance strategy, its processes, and the existing information technology. The data analysis part starts with a root cause analysis. Relevant data is acquired and preprocessed. Historical machine data, ERP maintenance data, and manual maintenance reports are therefore merged. Parameters, which can be associated with the behavior of the component, are identified. Using these parameters, different models are developed, and performance is evaluated. In addition, the IT infrastructure is modified and adapted to fit the requirements of PdM. As soon as both the model is developed and the infrastructure is available, personnel is trained, and the solution is deployed. Lastly, the company offers service packages from monitoring machine data to planning and controlling field workers and maintenance actions for the operation of the solution, which also includes continuous improvements.

The process is very structured; however, it allows for returns. In particular, a return to the functional design is often required to adapt the concept based on further findings. Three core competencies are named by the interviewee, which have to be present within a PdM team. These are engineering knowledge, data science knowledge, and functional knowledge (with regard to the maintenance process). The consultancy team consists of people with these three competencies.

Assessment and Adaptations: As for all consultancies, the process depicts a high level of maturity. It is in accordance with the developed process reference model. All phases are targeted. Nevertheless, three deviations are identified in the project preparation phase and adjusted in the process reference model. These are maturity assessment, functional design, and end-to-end use case definition. While maturity assessment and end-to-end use case definition are added lacking in the process reference model and are therefore included, functional design is subsumed under concept development as the same point is addressed.

Company J: Solution Provider + Management Consultancy

Process Description: Company J is an IT solution provider and management consultancy. They are offering PdM solutions, which make use of the developed standard software as a basis for development. However, a large part must be adapted to the customer's specificities. Their process is divided into three phases. Primarily, business analysis is carried out. Here, the current state is analyzed, and possibilities are elaborated, which are in accordance with the company's strategy. This step also includes a maturity assessment. Nevertheless, the interviewee outlines that the company's maturity is in many cases too low to be able to target PdM directly, and thus, data visualization is defined as the primary target. Given the target formulation, requirement analysis is performed from the business perspective. These are often defined by the customer and include different measures of success. This goes along with the use of case selection. Various possibilities are identified and prioritized subsequently. The use case with the highest priority is targeted in detail in the second phase. With regard to this specific use case, conceptual analysis is conducted. This includes the discussion of organizational aspects as well as the feasibility analysis of the selected use case in terms of existing data, data volumes, and technical infrastructure. Required sensor extensions are within the responsibility of the customer. With regard to the technical infrastructure, Company J offers different solutions, ranging from Hadoop clusters, relational databases, and cloud services. In case the feasibility analysis points out difficulties for the use case, a new case is selected. This is followed by the PHM model development. Here, several iterations of the CRISP-DM methodology are applied. Initially, models are chosen with a high degree of explainability and understandability, whereas in later iterations, more sophisticated models are considered. This is done to understand the data better and build confidence on the customer's side. The methodology starts with business understanding, in which the objectives for model development are refined, and the data mining problem is specified. Subsequently, data is collected and analyzed (data understanding) together with maintenance personnel and operators to gain further case knowledge. The selected data is harmonized, prepared, and mapped on the existing data schema of the standard solution. The solution provides basic functionalities, which can be adapted to the specific case. This can either be done using existing templates or by implementing methods from scratch. After model development, its performance is evaluated with regard to the requirements defined in business understanding. If performance is acceptable, further assets are targeted (re-execution of the second phase), and the last phase is initiated in terms of deployment. For this, the solution is embedded in the existing customer processes, and employees are trained. A dashboard is designed which offers visualizations to the customer and enables transparency of the decision support. Lastly, continuous improvement measures are implemented. This could either be self-adjusting machine learning procedures or regular manual adaptations.

The process is described as very agile with close feedback loops. Return to previous steps is possible; however, they are to be avoided to increase transparency for the customer. The CRISP-DM methodology is used, with sprints every 4-6 weeks. Besides experts from the company, the project team also required input from the production and maintenance department, the IT department as well as interface persons, which have data knowledge and are able to interpret the measured values. Furthermore, the support of the project by the business owners is of high importance.

Assessment and Adaptations: The process depicts a very high level of maturity. It is highly flexible for different PdM project characteristics. Considering the three phases, the process can be matched to the process reference model, which portrays a very similar structure. Nevertheless, the process led to minor extensions of the process reference model. As for Company I, maturity assessment has not been explicitly part of the model and therefore added to it. In addition, the conceptual analysis depicts the steps of operationalization validation and pre-feasibility analysis. Both steps have already been added by other case studies, and therefore only highlights the relevancy of these steps for practical realizations.

Company K: Solution Provider

Process Description: Company K implements a standard solution for prognostics. The purchase of the software also includes the support of the company during the installation and introduction for their customers. Nevertheless, the company is only focusing on the data analysis part of the process and thus requires a working system architecture. In the first step, requirements are formulated, and a machine is selected for an initial pilot. For this pilot machine, historical data is collected. These are primarily analyzed to identify failure causes and possible influencing factors. In addition, operator comments can be attached to the different identified failures. Thereafter, the solution is set up. A connection between the machine interfaces and the software is therefore established. Data is stored in a relational database. The solution provides a set of methods that learn automatically but require parameter adaptations. For this reason, parameters are configured afterward. The performance of each model is evaluated and matched against the initially defined requirements. If the performance is not sufficient, re-configuration is required. In parallel to this pilot phase, a cost-benefit analysis can be performed. While in the beginning, only analogies to other projects can be drawn, after model development, potential cost-saving are calculated in a retrospective manner. Following the pilot phase, further assets are included in the solution (similar to the process described above). In addition, interpretation training is provided to the customer. The dashboard provides decision-making support in terms of graphical visualizations and operator comments.

The project team combines experts from Company K and equipment experts from the customer, both operators and maintenance employees.

Assessment and Adaptations: The company has a very structured approach to realize PdM implementations for their customers. As they are providing a standard solution, the process is slightly different from the processes described above and to the process reference model. It mainly focuses on the setup and configuration of the prognostics solution. Nevertheless, there are strong similarities. In the described process, the PHM design phase is targeted as well as partly the project preparation phase and candidate solution deployment phase. A suitable system architecture is a pre-requisite that has to be fulfilled by the customer and is therefore not targeted by the company K. Deviations to the process reference model depict, on the one side, the parameter configuration and on the other side the dashboard interpretation trainings. The step parameter configuration is applicable only to the case of standard software. As the process reference model is rather targeting its own internal implementations, instead of the usage of standard software, it is not included in the model. Furthermore, the dashboard interpretation training is subsumed under personnel training and, therefore, does not depict a necessary adjustment.

G   Interview Step Analysis

Table 33 | Interview Step Analysis

Phase	Step	A	B	C	D	E	F	G	H	I	J	K
Preparation	As-Is Analysis		X					X	X		X	
	Maturity Assessment									X	X	
	Target Formulation/ Specification									X	X	
	Requirements Analysis	X	X					X	X	X	X	X
	Use Case Refinement						X					
	Cost-Benefit Analysis		X					X	X			X
	Assembly Group Definition			X								
	Maintenance Strategy Definition			X								
	Machine/ Candidate/ Use Case Identification	MS	MS	CI	CI	CI	UC	UC	MS	UC	UC	MS
	Team Formulation				X	X						
	Analytical Feasibility Assessment				X	X						
	Data Availability Assessment				X	X	X			X	X	
	Measurement Market Research Conduction	X										
	Measurement Technique Identification/ Testing/ Installation	X	X	X						X	X	
	Concept Development		X									
	Functional Design									X		
	End-To-End Use Case									X		
SA Design	Operationalization Validation							X		X	X	
	Market Analysis (Hardware)		X									
	Operationalization Concept							X	X			
	Technical Infrastructure		X					X	X	X	X	
PHM Design	Parameter/ Symptoms Identificat.			PI		SI						
	Software Solution	X		X								
	Solution Set-Up/ Parameter Configuration											X
	Data Acquisition	X	X	X			X			X	X	X
	Data Sighting						X	X	X	X		
	Data Preparation		X	X		X	X	X	X	X	X	
	Data Visualization	X	X									
	Anomaly Detection			X					X			
	Root-Cause Analysis/ Fault Cause Analysis	RC	FC		FC	FC				RC		X
	Fault Cluster Identification								X			
	Model/ Prototyp Development		(X)	(X)	PD	X	X	X	X	X	X	X
	Performance Evaluation			(X)	X		X	X	X	X	X	X
	Maintenance Instruction Definition					X						
	Solution Documentation				X	X						
	Visualization		X	(X)				X	X		X	X
Candidate Deployment	Industrial Implementation				X							
	Personnel Training	(X)				X				X		X
	Process Adjustments			(X)							X	
	Change Management			(X)						X		
	Field Testing					X						
	Deployment/ In-Service	(X)	(X)	(X)	IS	X	X	X	X	X	X	
	Sustainment Implementations	X			X	X		MR	X	X	X	
	Project Documentation			X			X					

H Algorithm Development Method for Fleet Prognostics

Table 34 | Advantages and Disadvantages of Fleet Strategies

	General Remarks/ Assumptions	Advantages	Disadvantages
Local Models	• Possible in cases with distinct groups of units • Accuracy highly depends on the selected method • Method selection based on the user's knowledge • Availability of failure distribution depends on the selected method • Model assumption	• Simple and intuitive/ understandability	• Not able to handle unseen behavior • Requires a large amount of data to build clusters with a reasonable amount of trajectories • Possibly a large number of models required • Does not target unit-to unit specification (general prediction model)
Similarity-Based Approach	• System future operation follows a similar pattern as the past[1] • The system is instrumented at discrete time[1] • Effect of the maintenance actions, if any, during the life cycle under consideration is negligible; otherwise, a life cycle is considered to be complete when major corrective maintenance action is carried out[1] • Catastrophic failures and infant failures are considered as exceptions for the period that RUL predictions are made[1] • The system under study shows an irreversible evolving degradation behavior[1] • Exploit similarity between an observed situation and cases already known and resolved[13]	• Very flexible, simple • No definition of failure threshold required[12] • Does not require fully understand the degradation behavior and employ the first-principle models[2] • No definition of model[1,3,5] • Non-stationary, non-parametric multi-variate approach[1,3] • Low computational load/ complexity during model training[1,3] • Continuous learning[3,4,5] • Long term accuracy[1,3,5,6,17] • Different initial degradation conditions[17] • Different degradation pattern[17]	• Requires large trajectory library[2,7,17] • High Computational requirements at runtime/ lazy learner [1,3] • Requires complete run-to-failure data • Large storage capacities • Not able to handle unseen behavior • No failure distribution • Performance highly dependent on the significance of library of reference scenario[5] • No consideration of covariates • Cases need to be well structured and described[13] • Uni-variate data required
Model-Based SBA		• Does not require work with a large library of previously recorded HI trajectories[4]	
Correction Model	• Data required and method capabilities depend on the selected method	• Simple, understandable, transparent • Able to handle "new" behavior that has not been seen beforehand	• No failure distribution • requires run-to-failure data • Computational demanding[5]

Bayesian Inference/Mixed Models	• Update probabilities of future observations by incorporating evidence from previous experience into the overall conclusion[10] • Data should be collected under similar use conditions and should reasonably span the range of individual variations between components[16]	• Handle uncertainty in a natural way (probability distribution)[10,11] • Handle right-censored data[9,16] • Natural hedge against overfitting[8,10] • Non-linear generalized model[9] • Dynamic update of model based on real-time data[9] • Model updating as more information is provided[8] • Identify pattern not seen within fleet[16] • Random level of initial degradation[16] • Low storage requirement[16] • For mixed models: simplicity, high accessibility[16]	• Requires model assumption[16]/ expert knowledge • Complex[8,10] • Requires failure threshold[16] • No explicit consideration of covariates • Very specific approach for a given scenario[11] • Requires substantial technical facility in advanced probability and statistics[8,10] • Computational expensive[8,10] • Low/ hard understanding[8,10] • Not applicable in cases with no physical model or not enough data to estimate conditional probabilities[5] • Requires run-to-failure data[16]
Bayesian Hierarchical M.	• Assumption: data vary from engine-to-engine, from build-to-build and from airline-to-airline or even finer levels of gradation[8]	• Two-stage regression (individual and group level)[8] • Different sets of covariates[8]	
Ensemble Learning	• Data required and method capabilities depend on the selected method	• More accurate RUL predictions when member algorithms producing diverse RUL predictions have comparable prediction accuracy[17] • Ensembles improve robustness and accuracy[17] • Inherent flexibility to incorporate any advanced prognostic algorithm that will be newly developed[17] • Short-run time[17]	• Black-box/ no understandability • High data requirement • Computational intensive training[17]
GPR		• Can use a small number of historical data sets to learn the degradation patterns[18] • Probability distribution[18] • Has high adaptability and is suitable for dealing with the RUL prediction issue of high-dimension and small-size data sets[19] • Can adapt to environments and learn from experience[20,21]	• Computational efforts (scales with the number of calculating data points for training $O(n3)$[18] • Requires failure threshold • Need to determine covariance function[20] • Only suitable for Gaussian likelihood[20]

Markov Models		• Handle right-censored data[22] • Can classify time-series data without specific knowledge of the problem, as long as sufficient data is available for training[14] • Provides the current health status and RUL related to operating conditions[22] • Multi-failure modes[23] • Quantifying uncertainty through variable operating conditions[22] • Multivariate signals[22]	• Neglects historical degradation behavior (Markov property)[14,22,24] • High data requirements for sHMMs[14] • Computationally intensive[14] • Separate model for each failure mode[14]
ANN		• Non-linear statistical data model, to model complex relationships[13]	• Behaves as blackbox[13]

Table 35 | Source Mapping for the Assessment of Fleet Prognostic Approaches

Number	Source
1	Wang 2010
2	Yu et al. 2012
3	Bleakie and Djurdjanovic 2013
4	Palau et al. 2020
5	Liu et al. 2007
6	Wang et al. 2008
7	Xue et al. 2008
8	Zaidan et al. 2015a
9	Wang et al. 2018
10	Zaidan et al. 2013
11	Ling and Mahadevan 2012
12	Bektas et al. 2019
13	ISO 13379
14	Sikorska et al. 2011
15	Lee et al. 2017
16	Coble and Hines Wesley 2011
17	Hu et al. 2012
18	Duong et al. 2018
19	Lei et al. 2018
20	Lee et al. 2014b
21	Atamuradov et al. 2017
22	Al-Dahidi et al. 2016
23	Kan et al. 2015
24	Si et al. 2011

Table 36 | Detailed Advantages and Disadvantages of Prognostic Methods

	Advantages	Disadvantages
Simple Trend Regression	• Random coefficient regression models can provide a PDF of the RUL[14] • Simplicity of their calculation and ease of implementation[3,5] • Curve fitting based methods are relatively simple to apply as they do not require a substantial amount of training data or a detailed physical model that describes the fault progression[6] • Easy to set alarms[3] • Simplicity and computationally efficiency can be performed in real-time[2,9]	• Assumption: underlying stability in the monitoring system; inaccurate in times of change assume monotonic trend[3,5,14] • Results are dependent on selecting an appropriate curve-fitting model form[6] • Simple thresholds are often unreliable for predicting remaining useful life, particularly where data needs to be extrapolated[3] • Not capable of handling noise well[3] • Availability of confidence limits dependent on the amount of data at the different states of failure development[3] • Interpretability is affected by process/ measurement noise and variations in operating conditions[3]
Autoregressive Models	• Accurate results for short-term prediction[3,9] • Historical failure data and failure mechanism are not required; a small amount of historical data[2,3,9] • Computationally efficient, can be performed in real-time[2,3,9] • Understanding of detailed failure mechanisms is not required[3]	• Less reliable when used for long-term predictions due to dynamic noise, their sensitivity to initial system conditions, and accumulation of systematic error in the predictor[3,4,9] • Application to linear time-invariant systems whose performance features display stationary behavior (advanced ARMA techniques for non-stationary data)[3,4,9] • No confidence limit[9] • Performance highly depends on the trend information of historical observations[2] • Application of the AR or ARMA models is difficult due to the complexity in modeling, especially the need to determine the order in the model[15] • Does not integrate prior or expert knowledge[3]
Stochastic Processes	• Distribution of FTP can be formulated analytically[14] • Required mathematical calculations are relatively straightforward, and the physical meaning is easy to understand[9,14]	• Wiener process for non-monotone deterioration (assumption: mean degradation path is linear); Gamma processes are adequate for monotone processes[9,14] • The variance of the noise term in the Wiener process is proportional to the time length is measured, which is a strong requirement not many state processes can possess[14]

| Markov Models | • Semi-Markov chains are more suitable in prognostics[9,16]
• HMM is also suitable for non-linear systems. HMM can estimate data distribution of normal operation with nonlinear and multimodal characteristics, assuming that predictable fault patterns are not available[1]
• Can be extended to non-stationary systems (approximated by a piece-wise stationary model)[1]
• causal relationships can be derived from incomplete or sparse data sets[3,9]
• Bayesian methods provide confidence limits, intrinsically[3,9]
• Models are also simple to construct (albeit often with expert guidance required), readily reusable, and easy to modify, adaptable to changes in knowledge bases or process configurations; HMMs are easy to interpret than ANNs[1,3]
• Efficient and robust modeling framework that efficiently captures and utilizes knowledge of domain experts that could otherwise be difficult to quantify with 'hard data'[1]
• Can be developed based on knowledge or learned with data; no knowledge of physical mechanism required if data is available[1]
• Able to model a number of different systems designs and failure scenarios[3,5]
• Be used to model systems rather than components, as the network lends itself to be easily configured hierarchically[3]
• Not only reflect the randomicity of machine behaviors but also reveal their hidden states, changing processes. Furthermore, HMMs have a well-constructed theoretical basis and easy to realize in software[16]
• Markov models are suitable in dealing with the multi-stage transition processes[2] | • Markovian properties, A Markov chain is a sequence of states where each depends only on the event immediately preceding; Markov assumption[1,2,3,7,14,16]
• If the number of states is unconstraint; severe overfitting might result (regularization, cross-validation)[17]
• Computational intensive (proportional to the number of states)[3]
• Basic model: Failure rate is constant (exception semi-MMs); it is assumed that the state transition matrix and the observation models do not change over time (this model only takes into account stationary or time-invariant parameters.)[1,9,17]
• Independence assumption: output observation at time t is dependent only on the current state; it is independent of previous observations and states (rarely valid in prognostics)[3,7,14,16]
• High amount of data required for (s)HMM[2,3,9,14]
• Separate models are required for each failure mode[3]
• Perfect repair assumed[3]
• Only the mean and variance of the RUL can be estimated so far, and it is difficult to obtain an explicit distribution of the RUL in a closed-form[14]
• Ultimately, a Bayesian network is only as useful as the prior knowledge is available and reliable; networks can be distorted and results invalidated by the exceedingly optimistic or pessimistic assignment of conditional probabilities. The quality of network prediction results will also be affected by the selection of a prior distribution. Therefore some understanding of the physical system is useful[3] |

Artificial Neural Networks	• Can model complex, multi-dimensional, unstable or non-linear systems[1,2,3,4] • High-capabilities in generalization, classification, and noise-immunity through non-linear information processing[1] • More capable of capturing and modeling complex phenomena without a priori knowledge; physical, no physical understanding required[1,3,5] • Remarkable ability in generalizing and deriving meaning from imprecise or complex data[1,2] • Can learn how to do tasks based on the training data or initial experience and create its own representation and organization of the information received during learning[1] • Fast computation and real-time operation due to parallel model structure[1,5] • Fast in handling multivariate analysis[1] • Fault tolerance behavior via redundant information coding enables the network to still retain some capabilities even with network damage[1] • Perform at least as good as the best traditional statistical methods without requiring untenable distributional assumptions[5] • ANN can estimate actual failure time with long prediction horizons when it is used for exponential projection or data interpolation[1] • Flexible with regard to the type of input data[3] • Adaptive system[4]	• Requires data preprocessing to limit the number of data inputs and reduce model complexity; pre-processing is critical for efficient and effective computation[1,3] • Requires a large amount of training data that have to be representative of true data range and its variability, complete run-to-failure data[1,2,3,6] • Operates as a black box without documentation of the qualitative information of the model; not clear about how decisions are reached in a trained network[1,2] • Model training is mostly trial and error and can therefore be time-consuming, no standard procedure[1,3,4] • Structure and parameters are initialized randomly/ manually, which reduces their generalization ability among different cases[2] • Most networks cannot provide confidence limits on the output(confidence prediction NNs)[3] • Outputs need to be mapped to a physical representation[3] • Requires sufficient computational resources[4] • Prone to overfitting[3]

Vector Machines	• Sparse technique (only requires support vectors to be stored), low storage requirements[7,8] • Computational complexity does not depend on the dimensionality of the input space[7,8,11] • Excellent generalization capability (maximized decision boundary), with high prediction accuracy[1,4,7,8] • Solution is found analytically rather than heuristically[7] • Known for their robustness, good generalization ability, and unique global optimum solutions[1,7,8,9] • Does not suffer from the course of dimensionality and overfitting, multilocal minima[7] • Automatically determines the model complexity by selecting the number of support vectors[7] • Excellent in processing multi-dimensional data[8] • SVM and RVM are superior to deal with the issues of small sample sizes[1,9] • Efficient for small or large data sets and real-time analysis[1,4]	• When trained with data that are not representative of the overall data population, hyperplanes are prone to poor generalization[7] • Standard SVM is computationally infeasible for large scale data[8] • No standard method to choose the kernel function, which is the key process for SVM; Parameters need to be specifically tuned for the problem at hand, and this may be difficult[1,2,9] • Lack of probabilistic output (SVM)[10]
Gaussian Process Regression	• GPR has high adaptability and is suitable for dealing with the RUL prediction issue of high-dimension and small-size data[1,2,12] • GP can be used with more flexibility for the non-linear, non-stationary regression problem; adapts to environments and learns from experience[1,2,4,12] • Easier to interpret and handle than their conventional counterparts[13] • GPR is a flexible, probabilistic non-parametric Bayesian model that allows a priori probability distribution to be defined over the space of functions directly[1] • Increased transparency[1] • Capture prior knowledge in the model[1] • Uncertainty estimates[1] • Allows non-parametric learning of a regression function from noisy data, avoiding simple parametric assumptions[1]	• Generally has heavy computational demand, especially when training data sets are large[1,2] • Difficult to find optimal values of the scale parameters, need to determine covariance function[1,4] • Only suitable for Gaussian likelihood[1,4] • Assumes all points are normally distributed, and error between every point is correlated[1]

Table 37 | Source Mapping for the Assessment of Prognostic Methods

Number	Source
1	Kan et al. 2015
2	Lei et al. 2018
3	Sikorska et al. 2011
4	Lee et al. 2014b
5	Heng et al. 2009
6	Lee et al. 2017
7	Awad and Khanna 2015
8	Huang et al. 2015
9	Xia et al. 2018
10	Goebel et al. 2008
11	ISO 13379
12	Sutharssan et al. 2015
13	Rasmussen and Williams 2008
14	Si et al. 2011
15	Jardine et al. 2006
16	Peng et al. 2010
17	Ghahramani 2001
18	Atamuradov et al. 2017
19	Wu et al. 2017a
20	Kim et al. 2017
21	Mathew et al. 2017
22	Singh et al. 2016
23	Kotsiantis et al. 2007
24	Tsui et al. 2015
25	Aizpurua and Catterson 2015
26	Saxena et al. 2009
27	Shalev-Shwartz and Ben-David 2014
28	Ye and Xie 2015
29	Bae and Kvam 2004

Table 38 | Data Preprocessing

	Step	Method	Source
Under-standing	Data Visualization	Sammon Mapping	(Peel 2008)
	Data Description	Various Data Statistics	(Heimes 2008)
Feature Engineering	Feature Extraction	Delta of operation parameters after fault initiation	(Xue et al. 2008; Coble and Hines Wesley 2011)
		Time-frequency distribution; time-shift invariant moments	(Liu et al. 2007)
		Root Mean Squared Value (vibration analysis)	(Wang et al. 2018)
	Sensor Selection	Grey Correlation Analysis to HI	(Yu et al. 2012)
		Empirical Observation (discrete values, non-monotonic trend, inconsistent trend)	(Yu et al. 2012; Wang et al. 2012; Wang et al. 2008; Wu et al. 2017b; Riad et al. 2010; Hu et al. 2012; Al-Dahidi et al. 2016)
		Experimental Analysis	(Ramasso 2014; Wang et al. 2008)
		Prognosis metrics: monotonicity, trendability, prognosability	(Bektas et al. 2019; Rigamonti et al. 2015; Coble and Hines Wesley 2011)
		Empirical Signal-Noise Ratio	(Wang 2010)
		Improved Distance Evaluation	(Bagheri et al. 2015)
	Dimensionality Reduction	Principal Component Analysis	(Liu et al. 2007; Wang 2010)
		Combined weighted average of all features	(Coble and Hines Wesley 2011)

Data Preparation	Normalization	Normalization	(Wu et al. 2017b)
	Smoothing	Simple Moving Average	(Riad et al. 2010; Yang et al. 2017; Ramasso 2014)
		Model fitting: Exponential fitting, Relevance Vector Machine, Support Vector Machine	(Wang et al. 2008; Hu et al. 2012)
	Dynamic Operating Conditions	Clustering operating variables	(Wang et al. 2008; Babu et al. 2016; Riad et al. 2010; Yu et al. 2012)
		Multi-regime normalization	(Bektas et al. 2019; Wang 2010; Babu et al. 2016; Riad et al. 2010; Peel 2008; Lim et al. 2014; Rigamonti et al. 2015; Al-Dahidi et al. 2016)
		HI calculation Per Regime and Fusion	(Yu et al. 2012; Ramasso 2014; Wang 2010; Palau et al. 2020; Wang et al. 2012; Hu et al. 2012; Wang et al. 2008)
		HI calculation with ANN (regime as input)	(Wang 2010; Bektas et al. 2019)
	Anomaly Detection	Binary Classifier	(Yang et al. 2017)
		Fleetwide populated causal tree	(Voisin et al. 2013)
		Tolerance Limits	(Zio and Di Maio 2010)
		Support Vector Machine	(Wu et al. 2017b)
		Elbow detection: z-test based method	(Rigamonti et al. 2015; Al-Dahidi et al. 2016)
		Quantization error	(Huang et al. 2007)
	Health State Estimation	Logistic Regression	(Li et al. 2015; Trilla et al. 2018; Bagheri et al. 2015)
		Linear Regression • learning only with HI = 0 & HI = 1 • learning with degradation data using an exponentially decreasing function	(Yu et al. 2012; Ramasso 2014; Wang 2010; Palau et al. 2020; Wang et al. 2012; Hu et al. 2012; Wang et al. 2008)
		Multi-layer perceptron	(Riad et al. 2010)
		MQE-based SOM	(Huang et al. 2007)
	Data Labeling	Linear decreasing Labels	(Peng et al. 2012; Wu et al. 2017b; Peel 2008; Lim et al. 2014)
		Piece-wise linear degradation	(Babu et al. 2016; Heimes 2008; Wu et al. 2017b; Peel 2008; Lim et al. 2014)
		Power or quadratic function	(Wu et al. 2017b)
		Negation of Remaining Cycles	(Wang et al. 2012; Wang et al. 2008)
Train-Test Split	Train-Test Split	Splitting into Train and Test Data	(Yang et al. 2017; Ramasso 2014; Wang 2010; Wu et al. 2017b; Gebraeel et al. 2004; Gebraeel and Lawley 2008; Al-Dahidi et al. 2016)

Table 39 | Performance Metrics

Class	Metric	Formula	Sources
Accuracy-based (Error-based)	Error (RUL residuals)	$$E_i = Y_i - \widehat{Y}_i$$	(Zaidan et al. 2013)
	Mean Absolute Error	$$MAE = \frac{1}{n}\sum_{i=1}^{n}|E_i|$$	(Ramasso 2014; Chang et al. 2017; Bektas et al. 2019; Bagheri et al. 2015; Wang et al. 2012) (Wang et al. 2018)
	Mean Squared Error / Root Mean Error	$$MSE = \frac{1}{n}\sum_{i=1}^{n}(E_i)^2$$	(Ramasso 2014; Bektas et al. 2019; Bleakie and Djurdjanovic 2013; Liu et al. 2007; Peng et al. 2012; Zio and Di Maio 2010; Peel 2008; Lim et al. 2014; Al-Dahidi et al. 2016; Wang et al. 2008)
	Mean Absolute Percentage Error / Mean relative error absolute percentage error	$$MAPE = \frac{100}{n}\sum_{i=1}^{n}\left|\frac{E_i}{Y_i}\right|$$	(Li et al. 2015; Ramasso 2014; Bektas et al. 2019; Guepie and Lecoeuche 2015; Al-Dahidi et al. 2016; Zio and Di Maio 2010; Chang et al. 2017)
	Anomaly Correlation Coefficient	$$ACC = \frac{\sum(\pi(i|j) - z_\#(i))(z_\# - z_\#(i))}{\sqrt{\sum(\pi(i|j) - z_\#(i))^2 \sum(z_\# - z_\#(i))^2}}$$	(Bektas et al. 2019)
	Bias	$$B = \frac{\sum_{i=P}^{n} E_i}{n - P + 1}$$	(Li et al. 2015)
Accuracy-based Metrics (Interval-based)	False Negative Rate / False Positive Rate	$$FP(i) = \begin{cases} 1 & (y_i\text{-}\widehat{y}_i) > t_{FP} \\ 0 & \text{otherwise} \end{cases}$$ $$FN(i) = \begin{cases} 1 & (y_i\text{-}\widehat{y}_i) > t_{FN} \\ 0 & \text{otherwise} \end{cases}$$ $$FPR = \frac{FP}{FP + FN}$$ $$FNR = \frac{FN}{FP + FN}$$	(Zaidan et al. 2013; Ramasso 2014; Bektas et al. 2019)
Precision-Based Metrics	Sample Standard Deviation / Precision / Steadiness Index	$$S_i = \sqrt{\frac{\sum_{i=1}^{n}(E_i - M_i)^2}{n - 1}}$$	(Bektas et al. 2019; Al-Dahidi et al. 2016; Rigamonti et al. 2015)
	Mean / Median Absolute Deviation from the Sample Median	$$AD_i = \frac{1}{n}\sum_{i=1}^{n}|E_i - median(E_i)|$$ $$AD_i = median(|E_i - median(E_i)|)$$	(Bektas et al. 2019)
Prognostic Performance	Prediction horizon	Chapter 5.4.2.3	(Yu et al. 2012; Wang 2010)
	Rate of acceptable predictions / $\alpha - \lambda$ accuracy / MA-α / α-Nmap	Chapter 5.4.2.3	(Yu et al. 2012; Wang 2010; Rigamonti et al. 2015; Wang et al. 2018; Al-Dahidi et al. 2016; Li et al. 2015)
	(Cumulative) Relative accuracy	Chapter 5.4.2.3	(Yu et al. 2012; Wang 2010) (Rigamonti et al. 2015) (Wang et al. 2018)

Convergence	Chapter 5.4.2.3	(Yu et al. 2012; Wang 2010; Al-Dahidi et al. 2016)
Performance score for parameter optimization	Chapter 5.4.2.3	(Wang 2010)
PHM Score	Chapter 5.4.2.3	(Ramasso 2014; Wang et al. 2012; Bagheri et al. 2015; Wang 2010; Babu et al. 2016; Wu et al. 2017b; Riad et al. 2010; Hu et al. 2012; Peel 2008; Lim et al. 2014; Coble and Hines Wesley 2011)

I   Experiments

Table 40 | Packages for Experiments

Packages	Description/ Purpose
OS.Path	Implements useful functions on pathnames
Panda	Data import, formatting, and manipulation
Numpy	Implements a variety of well-known algorithms (here: Euclidean distance, hierarchical clustering)
Scipy	Implements a variety of different methods
Tslearn/ Dtaidistance	Implements tools for time-series analysis (used: DTW)
Pyclustering	Implements clustering methods (used: k-medoids)
RPY2	Implements R-interface from Python
diptest (R)	Implements Diptest (R library)
Sklearn	Implements several Metrics (Silhouette coefficient)
Matplotlib	Used for Visualizations
Seaborn	Used for Visualizations
TS-Dist	Implements several distance measures (used: LCSS and EDR) – module available on Github (Too Yuen Min 2020)
DoE.base	Creates and supports full factorial experimental designs
Stats	Implements a variety of statistical methods (here: ANOVA)
Keras	High-level NN API Keras, used with Tensorflow in the backend

Bisher erschienene und geplante Bände der Reihe
Advances in Information Systems and Management Science

ISSN 1611-3101

Bd. 1: Lars H. Ehlers
Content Management Anwendungen.
Spezifikation von Internet-Anwendungen
auf Basis von Content Management Systemen

ISBN 978-3-8325-0145-7   40.50 €
285 Seiten, 2003

Bd. 2: Stefan Neumann
Workflow-Anwendungen
in technischen Dienstleistungen.
Eine Referenz-Architektur für die Koordination
von Prozessen im Gebäude- und Anlagenmanagement

ISBN 978-3-8325-0156-3   40.50 €
310 Seiten, 2003

Bd. 3: Christian Probst
Referenzmodell für IT-Service-Informationssysteme.

ISBN 3-8325-0161-4   40.50 €
315 Seiten, 2003

Bd. 4: Jan vom Brocke
Referenzmodellierung.
Gestaltung und Verteilung von Konstruktionsprozessen

ISBN 978-3-8325-0161-7   40.50 €
424 Seiten, 2003

Bd. 5: Holger Hansmann
Architekturen Workflow-gestützter PPS-Systeme.
Referenzmodelle für die Koordination von Prozessen
der Auftragsabwicklung von Einzel- und Kleinserienfertigern

ISBN 978-3-8325-0282-9   40.50 €
299 Seiten, 2003

Bd. 6: Michael zur Muehlen
Workflow-based Process Controlling.
Foundation, Design, and Application of workflow-driven Process
Information Systems

ISBN 978-3-8325-0388-8   40.50 €
315 Seiten, 2004